MASTER MECHANICAL APTITUDE & SPATIAL SKILLS

Practice Workbook with 300+ Questions

Copyright © 2021 by Complete Test Preparation Inc. ALL RIGHTS RESERVED. No part of this book may be reproduced or transferred in any form or by any means, graphic, electronic, or mechanical, including photocopying, recording, web distribution, taping, or by any information storage retrieval system, without the written permission of the author.

Notice: Complete Test Preparation Inc. makes every reasonable effort to obtain from reliable sources accurate, complete, and timely information about the tests covered in this book. Nevertheless, changes can be made in the tests or the administration of the tests at any time and Complete Test Preparation Inc. makes no representation or warranty, either expressed or implied as to the accuracy, timeliness, or completeness of the information contained in this book. Complete Test Preparation Inc. makes no representations or warranties of any kind, express or implied, about the completeness, accuracy, reliability, suitability or availability with respect to the information contained in this document for any purpose. Any reliance you place on such information is therefore strictly at your own risk.

The author(s) shall not be liable for any loss incurred as a consequence of the use and application, directly or indirectly, of any information presented in this work. Sold with the understanding, the author(s) is not engaged in rendering professional services or advice. If advice or expert assistance is required, the services of a competent professional should be sought.

The company, product and service names used in this publication are for identification purposes only. All trademarks and registered trademarks are the property of their respective owners. Complete Test Preparation Inc. is not affiliated with any educational institution.

Complete Test Preparation Inc. is not affiliated with the International Brotherhood of Electrical Workers, who are not involved in the production of, and do not endorse this publication.

We strongly recommend that students check with exam providers for up-to-date information regarding test content.

Version 8.5 June 2025
Published by
Complete Test Preparation Inc.
Victoria BC Canada
Visit us on the web at https://www.test-preparation.ca
Printed in the USA
ISBN-13: 978-1-77245-518-2

About Complete Test Preparation Inc.

Why Us?
The Complete Test Preparation Team has been publishing high quality study materials since 2005, with a catalog of over 145 titles, in English, French, Spanish and Chinese, as well as ESL curriculum for all levels.

To keep up with the industry changes we update everything all the time!

And the best part?
With every purchase, you're helping people all over the world improve themselves and their education. So thank you in advance for supporting this mission with us! Together, we are truly making a difference in the lives of those often forgotten by the system.

Charities that we support -
https://www.test-preparation.ca/charities-and-non-profits/

You have definitely come to the right place.
If you want to spend your valuable study time where it will help you the most - we've got you covered today and tomorrow.

Feedback

We welcome your feedback. Email us at feedback@test-preparation.ca with your comments and suggestions. We carefully review all suggestions and often incorporate reader suggestions into upcoming versions. As a Print on Demand Publisher, we update our products frequently.

CONTENTS

6 GETTING STARTED

8 MECHANICAL COMPREHENSION PRACTICE

Simple Machines	9
Answer Key	33
Circuits and Electricity	45
Answer Key	76
Basic Physics	94
Answer Key	108

115 SPATIAL RELATIONS

Folding and Rotating	116
Answer Key	131
Assembly	133
Answer Key	142
Line Following	145
Answer Key	154
Touching Blocks	160
Answer Key	168
Blocks	170
Answer Key	177
Cut Outs	180
Answer Key	189
Jigsaw	191
Answer Key	200

Matching Shapes	203
Answer Key	211
Visual Comparison	213
Answer Key	219

221 CONCLUSION

Getting Started

CONGRATULATIONS! By deciding to prepare for the Mechanical Comprehension test, you have taken the first step toward acing the test! Of course, there is no point in taking this important examination unless you intend to do your best to earn the highest grade you possibly can. That means getting yourself organized and discovering the best approaches, methods and strategies to master the material. Yes, that will require real effort and dedication, but if you are willing to focus your energy and devote the study time necessary, before you know it you will be on your way to a brighter future!

The Questions

The questions are centered on the basic principles in Engineering and Mechanical concepts. The practice questions are designed to assess your understanding and command of basic principles. Here are the most common subject areas:

- Pulleys and Belts
- Gears
- Springs
- Levers
- Acceleration
- Magnetism
- Horseshoe magnets

- Conductors
- Acceleration
- Open and closed circuits
- Switches
- Series and parallel circuits
- Electrical load and path
- Basic Physics

Spatial Relations

- Assembly
- Blocks
- Cut out
- Folding
- Rotation
- Touching blocks
- Matching
- Visual comparison
- Jigsaw
- Line following

PRACTICE QUESTIONS

Simple Machines

	A	B	C	D	E			A	B	C	D	E
1	○	○	○	○	○		31	○	○	○	○	○
2	○	○	○	○	○		32	○	○	○	○	○
3	○	○	○	○	○		33	○	○	○	○	○
4	○	○	○	○	○		34	○	○	○	○	○
5	○	○	○	○	○		35	○	○	○	○	○
6	○	○	○	○	○		36	○	○	○	○	○
7	○	○	○	○	○		37	○	○	○	○	○
8	○	○	○	○	○		38	○	○	○	○	○
9	○	○	○	○	○		39	○	○	○	○	○
10	○	○	○	○	○		40	○	○	○	○	○
11	○	○	○	○	○		41	○	○	○	○	○
12	○	○	○	○	○		42	○	○	○	○	○
13	○	○	○	○	○		43	○	○	○	○	○
14	○	○	○	○	○		44	○	○	○	○	○
15	○	○	○	○	○		45	○	○	○	○	○
16	○	○	○	○	○		46	○	○	○	○	○
17	○	○	○	○	○		47	○	○	○	○	○
18	○	○	○	○	○		48	○	○	○	○	○
19	○	○	○	○	○		49	○	○	○	○	○
20	○	○	○	○	○		50	○	○	○	○	○
21	○	○	○	○	○		51	○	○	○	○	○
22	○	○	○	○	○		52	○	○	○	○	○
23	○	○	○	○	○		53	○	○	○	○	○
24	○	○	○	○	○		54	○	○	○	○	○
25	○	○	○	○	○		55	○	○	○	○	○
26	○	○	○	○	○		56	○	○	○	○	○
27	○	○	○	○	○		57	○	○	○	○	○
28	○	○	○	○	○							
29	○	○	○	○	○							
30	○	○	○	○	○							

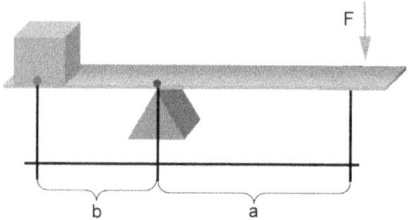

1. Consider the illustration above and the corresponding data:

Weight = W = 200 pounds
Distance from fulcrum to Weight = b = 10 feet
Distance from fulcrum to point where force is applied = a = 20 feet
How much force (F) must be applied to lift the weight?

   a. 80
   b. 100
   c. 150
   d. 200

2. A force of 20 kg. is applied to two springs in series, which compresses the springs 6 inches. If the same force is applied to springs in parallel, how far will the springs compress?

   a. 6 inches
   b. 3 inches
   c. 2 inches
   d. 1 inch

3. You are asked to determine the gear ratio of a vehicle. You open the differential and observe the ring gear the and pinion gear. The ring gear has 40 teeth and the pinion gear has 8, What is the gear ratio of the vehicle?

    a. 4:1
    b. 5:1
    c. 8:2
    d. 8:0

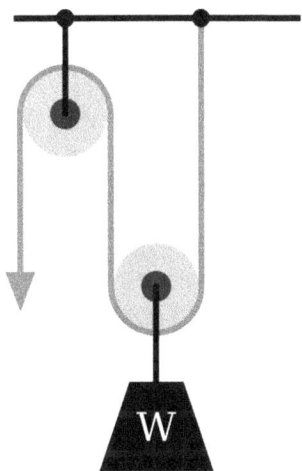

4. Consider the pulley arrangement above. If the weight, W, is 50 pounds, how much force is required to lift it?

    a. 10 pounds
    b. 20 pounds
    c. 25 pounds
    d. 50 pounds

5. Consider a gear train with 3 gears, from left to right, A with 20 teeth, gear B with 60 teeth, and gear C with 10 teeth. Gear A turns clockwise at 60 rpm. What direction and speed in rpm does Gear C turn?

    a. 120 rpm, clockwise

    b. 100 rpm clockwise

    c. 120 rpm counter clockwise

    d. 140 rpm counter clockwise

6. If a 100-pound object is sitting on a 10-square-inch plate, what is the PSI?

    a. 5

    b. 10

    c. 15

    d. 20

7. What is mechanical advantage?

    a. The ratio of energy input to energy output, typically where the input is less than the output.

    b. The ratio of energy input to energy output, typically where the input is greater than the output.

    c. The ratio of energy resistance to energy output, typically where the resistance is less than the output.

    d. None of the above

8. What is the ratio of mechanical advantage of a simple pulley?

    a. 2:1

    b. 1:1

    c. 3:1

    d. 1:2

9. Consider moving an object with a lever and a fulcrum. What is the relationship between the distance from the fulcrum and the speed the object will move?

    a. The farther away from the fulcrum, the faster the object will move.

    b. The closer to the fulcrum, the faster an object will move.

    c. An object will move the fastest when directly above the fulcrum.

    d. None of the above.

10. Which of the following are examples of a wedge?

    a. Corkscrew

    b. Scissors

    c. Wheelbarrow

    d. Pulley

11. Which of the following illustrates the principal of the lever?

    a. The greater the distance over which the force is applied, the greater the force required (to lift the load).

    b. The greater the distance over which the force is applied, the smaller the force required (to lift the load).

    c. The smaller the distance over which the force is applied, the smaller the force required (to lift the load).

    d. None of the above

12. Consider two gears on separate shafts that mesh. The input gear has 30 teeth and turns at 100 rpm. If the output gear has 40 teeth, how fast is the output gear turning?

    a. 300 rpm

    b. 250 rpm

    c. 75 rpm

    d. 100 rpm

13. Consider two gears on separate shafts that mesh. The input gear has 100 teeth and turns at 50 rpm. If the output gear has 20 teeth, how fast is the output gear turning?

    a. 300 rpm
    b. 250 rpm
    c. 200 rpm
    d. 100 rpm

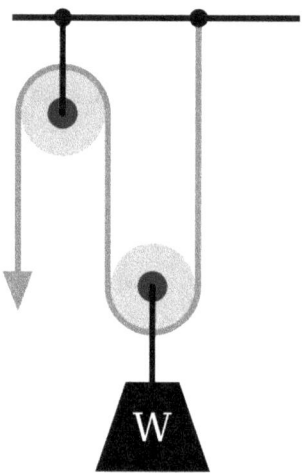

14. Consider the pulley arrangement above. If the weight is 100 pounds, how much force is required to lift it?

    a. 20 pounds
    b. 33 pounds
    c. 50 pounds
    d. 75 pounds

15. Tension of 40 kg. is applied to two springs in parallel, which expands the springs 8 inches. If the same force is applied to springs in series, how far will the springs expand?

   a. 2 inches
   b. 4 inches
   c. 8 inches
   d. 16 inches

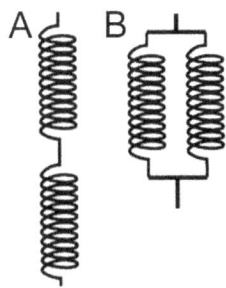

16. Consider the diagram above and select the correct labels from the options below.

   a. A - series, B - parallel
   b. A - parallel, B - series
   c. Series and parallel do not apply to springs
   d. None of the above

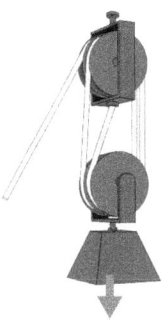

17. Consider the pulley arrangement above. If the weight is 200 pounds, how much force must be exerted downward on the rope?

   a. 200 pounds
   b. 100 pounds
   c. 50 pounds
   d. 25 pounds

18. Up-and-down or back-and-forth motion is called:

   a. Rotary motion
   b. Reciprocating motion
   c. Agitation motion
   d. Harmonic motion

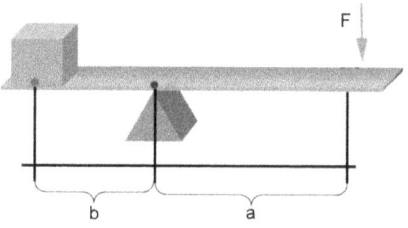

19. Consider the illustration above and the corresponding data:
Weight = W = 80 pounds
Distance from fulcrum to Weight = b = 10 feet
Distance from fulcrum to point where force is applied = a = 20 feet
How much force (F) must be applied to lift the weight?

   a. 80
   b. 40
   c. 20
   d. 10

20. The output torque of a 2 gear train is 1,000 newton-meters, and the gear ratio is 2:1. What is the input force?

   a. 200
   b. 400
   c. 500
   d. 1000

21. Which of the following is an example of torque?

   a. The wheel of a pulley turning
   b. A piston moving
   c. A horse pulling a load
   d. A tow truck pulling a vehicle

22. Find the weight of load L in N, if the pulling force F = 20N.

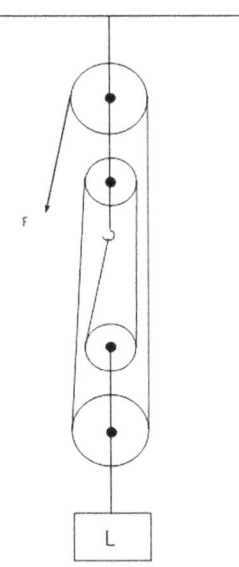

a. 5
b. 100
c. 20
d. 80

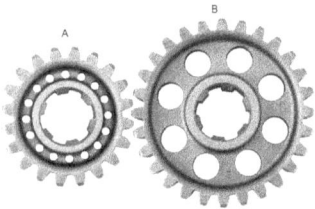

23. How many turns does the gear B make when the gear A makes 14 complete turns?

    a. 8
    b. 10
    c. 20
    d. 28

24. Which of the following is true about the system of meshed gears shown?

a. Gear A rotates slower than gear B
b. Gear A rotates slower than gear C
c. Gear B rotates slower than gear C
d. Gear B rotates faster than the other two gears

25. Which of the following is true of the relationship between screws and threads?

a. The larger the distance between threads, the easier to turn.

b. The smaller the distance between threads, the easier to turn.

c. The smaller the distance between threads, the more difficult to turn.

d. None of the above

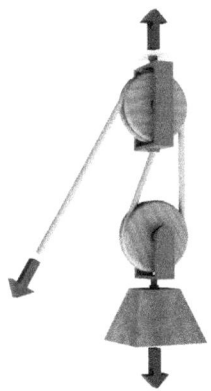

26. Consider the arrangement of pulleys above. If the weight shown is 150 pounds, how much force much be exerted to lift the weight?

a. 150 pounds
b. 100 pounds
c. 75 pounds
d. 50 pounds

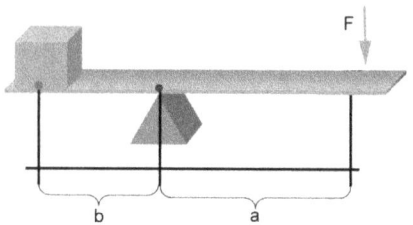

27. Consider the illustration above and the corresponding data:

Weight = W = 100 pounds
Distance from fulcrum to Weight = b = 5 feet
Distance from fulcrum to point where force is applied = a = 10 feet
How much force (F) must be applied to lift the weight?

    a. 100
    b. 50
    c. 25
    d. 10

28. Consider a gear train with 3 gears, from left to right, A with 10 teeth, gear B with 40 teeth, and gear C with 10 teeth. Gear A turns clockwise at 80 rpm. What direction and speed in rpm does Gear C turn?

    a. 100 rpm, clockwise
    b. 80 rpm clockwise
    c. 120 rpm counter clockwise
    d. 100 rpm counter clockwise

29. A force of 40 kg. is applied to two springs in parallel, which compresses the springs 10 inches. If the same force is applied to springs in series, how far will the springs compress?

   a. 40 inches
   b. 5 inches
   c. 10 inches
   d. 5 inches

30. Tension of 40 kg. is applied to two springs in series, which expand the springs 20 inches. If the same amount of tension is applied to springs in parallel, how far will the springs expand?

   a. 20 inches
   b. 10 inches
   c. 5 inches
   d. 2 inch

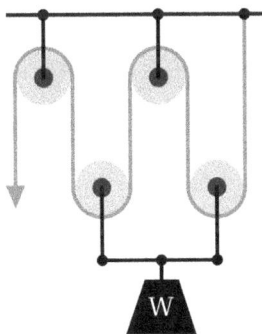

31. Consider the pulley arrangement above. If the weight, W, is 100 pounds, then how much force is required to lift the weight?

   a. 100 pounds
   b. 50 pounds
   c. 25 pounds
   d. 20 pounds

32. A cam is a mechanical linkage that:

a. Transforms linear motion into rotary motion and vice versa.

b. Transforms oscillating motion in to linear motion and vice versa.

c. Transforms reciprocating motion to oscillating motion.

d. None of the above

33. What is the function of the crankshaft?

a. To transform the back-and-forth motion of the pistons into rotary motion.

b. To transform rotary motion into reciprocal motion.

c. To transfer the rotary motion of the cam to the wheels

d. None of the above.

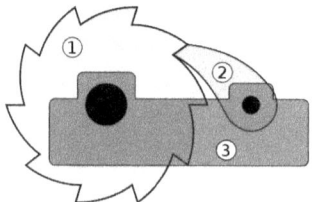

34. Identify the components labeled above.

a. 1 - ratchet, 2 - pawl, 3 - base

b. 1 - pawl, 2 - ratchet, 3 - base

c. 1 - gear, 2 - stop, 3 base

d. None of the above

35. Which equation below shows the relationship between F and P for the system of pulleys shown?

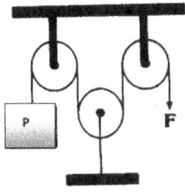

a. P = 3F
b. P = 2F
c. P = F
d. P = F/2

36. How many newtons of force is needed to pull the object up an inclined plane, if the weight of the object is 200 N?

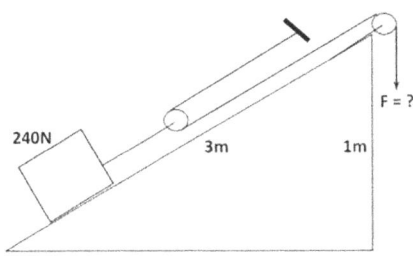

a. 50
b. 100
c. 150
d. 200

37. What is the force applied to lift the 400 N weight shown?

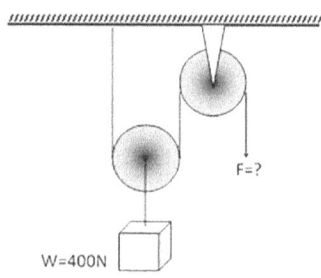

a. 200 N
b. 300 N
c. 400 N
d. 800 N

38. How many turns does gear 1 make when gear 3 makes 210 turns?

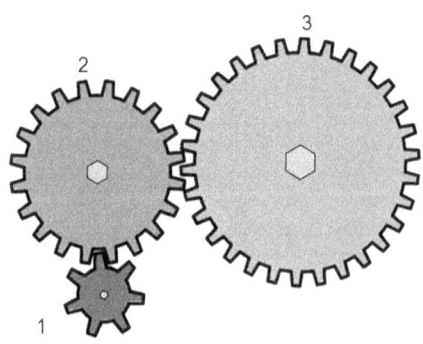

a. 30
b. 90
c. 300
d. 900

39. What is the minimum force (in N) needed to lift the 600 N object, if x = 3 m and y = 2 m?

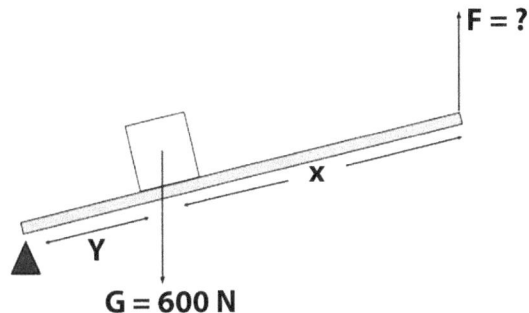

a. 400
b. 320
c. 240
d. 180

40. What is the distance between the mass m and the fulcrum, if the system is in equilibrium and the length of the rod d is 120 cm? Give the answer in cm.

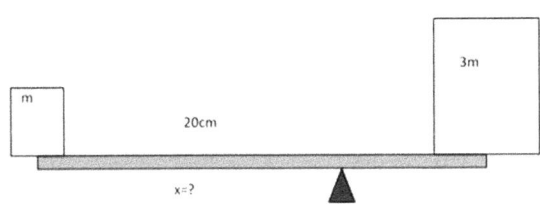

a. 80
b. 85
c. 90
d. 95

41. What is the value of the force F enough to lift the object up, if the weight W of the object is 360 N?

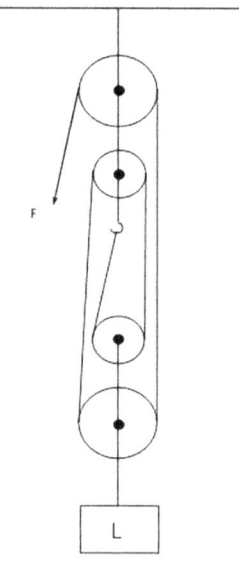

a. 180 N
b. 120 N
c. 90 N
d. 72 N

42. How many turns does the gear A make if the gear B makes 100 turns?

a. 175
b. 70
c. 7
d. 5

43. Which of the following statements about gears is false?

   a. Gears are teethed wheels used to generate rotation
   b. Meshed gears move at the same time
   c. Meshed gears move at the same speed
   d. Teeth in gears help increase the friction and avoid slipping

44. What direction does the rack in the figure move if the pinion rotates clockwise?

   a. Left
   b. Right
   c. Up
   d. Down

45. What direction does the pinion in the figure rotate if the rack shifts on the right?

   a. Clockwise
   b. Counterclockwise
   c. First clockwise then counterclockwise
   d. First counterclockwise, then clockwise

46. The system shown is in equilibrium and the rod is weightless. What is the ratio P/F ?

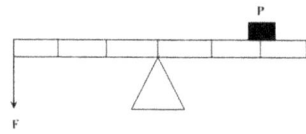

a. 3/2
b. 2/3
c. 1
d. 5/3

47. What is the ratio of the load to effort?

a. Torque
b. Mechanical Advantage
c. Energy
d. Mechanical Energy

48. Which type of lever does the wheel and axle system shown represent?

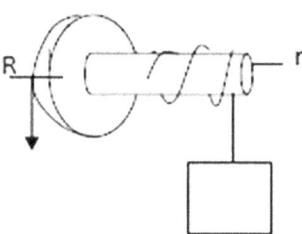

a. First class lever
b. Second class lever
c. Third class lever
d. Fourth class lever

49. What is the working principle of sugar tongs?

 a. First class lever
 b. Second class lever
 c. Third class lever
 d. Fourth class lever

50. Find the value of the ratio F/W, if R/r = 3.

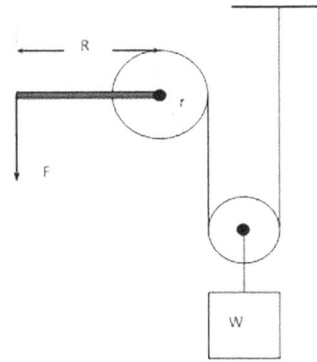

 a. 3
 b. 1/3
 c. 6
 d. 1/6

51. In the figure above m_1 = 4 kg and m_2 = 2 kg. What is the value of m_3 in kg if the system is in equilibrium? (The rod is weightless)

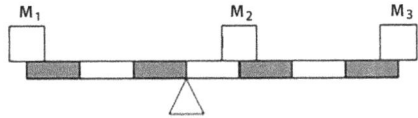

 a. 2
 b. 2.5
 c. 3
 d. 3.5

52. What is the reading of dynamo-meter in the figure below if the system is in equilibrium? The bar has no mass.

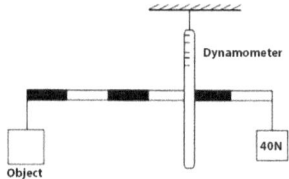

a. 20N
b. 40N
c. 60N
d. 80N

53. A uniform rod can be hold in equilibrium with the help of a system of pulleys as in the figure below. What is the weight of the rod if the force F = 3N?

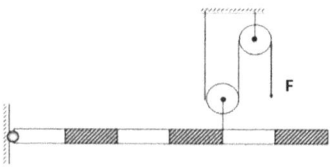

a. 3N
b. 4N
c. 6N
d. 8N

54. The length of the lever is 1 meter. What is the mechanical advantage of the system?

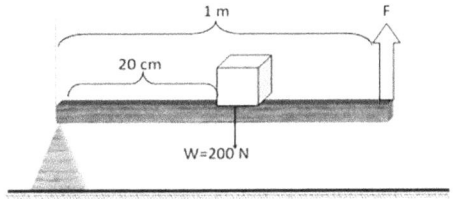

a. 1/5
b. 5
c. 1/2
d. 2

55. What is the effort-distance for the system shown?

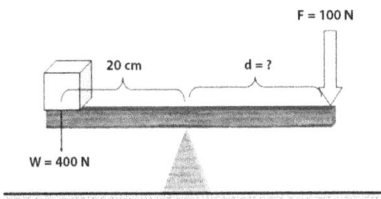

a. 5 cm
b. 40 cm
c. 80 cm
d. 100 cm

56. A door handle is an example of

   a. Inclined plane
   b. Pulley
   c. Screw
   d. Lever

57. The values of the acting force and the load are F = 20 N and P = 100 N respectively for the system shown. The total length of the lever rod is 120 cm. What is the distance between the force and the fulcrum (support) in cm?

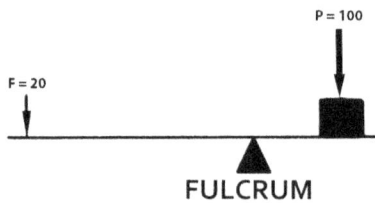

   a. 70
   b. 80
   c. 90
   d. 100

Answer Key

1. B
To solve for F, Weight X b (distance from fulcrum to weight) = Force X a (distance from fulcrum to point where force is applied)
200 X 10 = F X 20
2000/20 = F
F = 100

2. B
If the springs in series compress 6 inches, then the springs in parallel will compress half that amount, or 3 inches.

3. B
Opening the differential, the ring gear is the larger gear and the pinion the smaller. The gear differential is calculated by dividing the number of teeth on the pinion into the number of teeth on the ring gear. 40/8 = 5, or 5:1.

4. C
Since the weight is only attached to one pulley, the force required will be 50/2 = 25 pounds.

5. A
First calculate the speed of gear B. The gear ratio is 60:20 or 3:1. If gear A is turning at 60 rpm, then gear B will turn at 30/3 = 20 rpm.

Next calculate B and C. Gear C is smaller, so it will turn faster. The gear ratio is 60:10 or 6:1, and since gear B turns at 20 rpm, gear C will turn at 20 X 6 = 120 rpm.

Next calculate the direction. Gear A is turning clockwise, so Gear B is turning counter-clockwise, so gear C must be turning clockwise.

6. B
Calculate the PSI by taking the weight divided by the size of the object the weight is bearing on. 100/10 = 10 PSI.

7. A
Mechanical advantage is the ratio of energy input to energy output, typically where the input is less than the output. Mechanical advantage is a measure of the force amplification achieved by using a tool, mechanical device or machine system. Ideally, the device preserves the input power and simply trades off forces against movement to obtain a desired amplification in the output force. The model for this is the law of the lever. Machine components designed to manage forces and movement in this way are called mechanisms.

8. B
The ratio of mechanical advantage of a simple pulley is 1:1.

9. A
The farther away from the fulcrum, the faster the object will move.

10. B
Examples of wedges include the cutting edge of scissors, knives, screwdrivers, doorstops, nails axes and chisels.

11. B
The greater the distance over which the force is applied, the smaller the force required (to lift the load).

12. C
Call the input gear G^1 and the output gear G^2. Call the speed of G^1, S^1 and the number of teeth T^1. Similarly for G^2, we have S^2 and T^2.
Given data:
$S^1 = 100$
$T^1 = 30$
$S^2 = $ unknown
$T^2 = 40$
We know that $S^1 \times T^1 = S^2 \times T^2$
So, $100 \times 30 = S^2 \times 40$
$S^2 = 3000/40 = 75$ rpm

13. B
Call the input gear G^1 and the output gear G^2. Call the speed of G^1, S^1 and the number of teeth T^1. Similarly for G^2, we have S^2 and T^2.

Given data
S^1 - 50
$T^1 = 100$
S^2 = unknown
$T^2 = 20$
We know that $S^1 \times T^1 = S^2 \times T^2$

So, $50 \times 100 = S^2 \times 20$
$S^2 = 5000/20 = 250$ rpm

14. B
Notice the weight is attached to one end of the rope and to one pulley. The force required to lift a 100 pound weight with this arrangement is 100/3 = 33.

15. A
If the springs in parallel expand 10 inches, then the springs in series will expand twice that amount, or 20 inches.

16. A
The correct labels are, A - series, B - parallel

17. C
50 pounds of force much be exerted downward on the rope to lift the 200 pound weight. Since there are 4 pulleys, each will take 1/4 of the load. 200/4 = 50 pounds.

18. B
Up-and-down or back-and-forth motion is called reciprocal motion.

19. B
To solve for F, Weight X b (distance from fulcrum to weight) = Force X a (distance from fulcrum to point where force is applied)
$80 \times 10 = F \times 20$
$800/20 = F$
$F = 40$

20. C
If the output force is 1,000 newton-meters, and the gear ration is 2:1, the input force will be 1,000/2 = 500.

21. A
The wheel of a pulley turning is an example of torque. Torque, is the tendency of a force to rotate an object about an axis, fulcrum, or pivot. Just as a force is a push or a pull, a torque can be thought of as a twist to an object.

22. D
The block and tackle system composed of a system of pulleys as shown operates according the following rule:

Pulling Force=Load/(Number of supporting ropes)
Here, the number of supporting ropes is 4. So, we have
20 = Load/4
So, Load = 20 × 4 = 80 N.

Do not confuse the number of supporting ropes. The rope, which is being pulled, is not counted. Otherwise, you will obtain the wrong answer, Choice B 100 (20 × 5).

23. B
The equation of meshed gears states that the speed of rotation V (in rot/s) is inversely proportional to the number of teeth N. Mathematically,

$$V_A/V_B = N_B/N_A$$

From the figure, it is obvious that N_A = 20 and N_B = 28. So, we have

$$14/V_B = 28/20$$

$$V_B = (14 \times 20)/28 = 10 \text{ turns}$$

24. C
In meshed gears, larger the gear, slower the rotation and vice versa. Thus, gear B rotates slower than the others and gear A rotates the fastest.

25. B
The smaller the distance between threads, the easier to turn.

26. C
75 pounds of force much be exerted downward on the rope to lift the 150 pound weight.

27. B
To solve for F, Weight X b (distance from fulcrum to weight) = Force X a (distance from fulcrum to point where force is applied)
100 X 5 = F X 10
500/10 = F
F =50

28. B
First calculate the speed of gear B. The gear ratio is 10:40 or 1:4. If gear A is turning at 80 rpm, then gear B, which is larger, will turn slower, 80/4 = 20 rpm.

Next calculate B and C. Gear C is smaller, so it will turn faster. The gear ratio is 40:10 or 4:1, and since gear B turns at 20 rpm, gear C will turn at 20 X 4 = 80 rpm.

Next calculate the direction. Gear A is turning clockwise, so Gear B is turning counter clockwise, so Gear C must be turning clockwise.

29. B
If the springs in parallel compress 10 inches, then the springs in series will expand half that amount, or 5 inches.

30. B
If the springs in parallel expand 20 inches, then the springs in series will expand twice that amount, or 10 inches.

31. C
Notice the weight is attached to two of the pulleys. The weight required will therefore be 100/4 = 25 pounds.

32. B
A cam is a rotating or sliding piece in a mechanical linkage used especially in transforming rotary motion into linear motion or vice-versa

33. A
The function of the crankshaft is to transform the back-and-forth motion of the pistons into rotary motion.

34. A
The labeled components are, 1 - ratchet, 2 - pawl, 3 - base.

35. C
Here, there are 3 fixed pulleys forming a single system. It is known that fixed pulleys do not provide any gain in force. So, we have P = F

36. A
Here we have the combination of two systems composed of an inclined plane and a movable pulley.
The equation of the inclined plane is
Load/Force=(Path distance)/Height=Mechanical advantage

So, the mechanical advantage MA of the inclined plane is
MA = 2h/h = 2
The equation of the movable pulley is
Mechanical advantage= (Load)/(Force) = 2

Therefore, the total mechanical advantage is 2 × 2 = 4. This means the force needed to lift the 200 N weight is 200N / 4 = 50N (choice A).

37. A
Here, we have a system of combined pulley, i.e. one fixed and one movable. The fixed pulley does not provide any mechanical advantage while the movable pulley provides a mechanical advantage of 2.
So, the force required to lift the 400 N weight (load) is:
F = W/2 = 400N/2 = 200N

38. D
The equation of meshed gears states that the speed of rotation V (in rot/s) is inversely proportional to the number of teeth N. Mathematically,

$N_1 \cdot V_1 = N_2 \cdot V_2 = N_3 \cdot V_3$

Here, we are concerned only for the gears 1 and 3. Thus, we have

$7 \cdot V_1 = 30 \cdot 210$
$V_1 = (30 \cdot 210)/7 = 900$ turns

39. C

This is a second-class lever as the Load is between pivot and force.

The equation of levers is
Load × Load distance = Force × Force distance
Here, Load = G = 600N, Load distance = y = 2m, Force distance = x + y = 3m + 2m = 5m and calculate the force.
So,

600 X 2 = F X 5

F = (600 X 2)/5 = 1200/5 = 240N

40. C

This is an example of first-class lever as the pivot (fulcrum) is between Load and Force.

The equation of levers is
Load × Load distance = Force × Force distance

Here, Load = 3m · g, Force = m · g, Load distance = (120 – x) cm, and Force distance = x cm. Here, we have to calculate the force distance. So,

3mg × (120 - x) = mg × x

Simplifying mg from both sides, the equation becomes,

3 × (120 - x) = x

3 × 120 – 3 × x = x

4x = 360

X = 90cm

41. C

The block and tackle system composed of a system of pulleys as shown operates according the following rule:

Pulling Force = Load / (Number of supporting ropes)

Here Load and Weight are the same thing.

Here, the number of supporting ropes is 4. So, we have

F = 360/4

Force = 90 N Choice C

Feedback for Choice D

Do not confuse the number of supporting ropes. The rope which is being pulled is not counted. Otherwise, you will obtain the wrong answer choice D 72 (360 / 5).

42. A

The equation of meshed gears states that the speed of rotation V (in rot/s) is inversely proportional to the number of teeth N. Mathematically,
VA/VB = NB/NA

From the figure, it is obvious that NA = 20 and NB = 35. So, since the time of rotation for both gears is equal, we have

VA/100 = 35/20

VB = (100 × 35)/20 = 175 turns

43. C

Choice A is correct. Gears are teethed wheels used to generate rotation.
Choice B is correct. Meshed gears move at the same time as they are connected.
Choice C is false. Meshed gears move at different speed depending on their size. Larger gears move slower than smaller gears.
Choice D is correct. Teeth in gears help increase the friction and avoid slipping.

44. A
If the pinion rotates clockwise, its lower teeth move due left. Therefore, the rack shifts left as well.

45. B
If the rack shifts to the right, the lower part of the pinion moves right as well. This means the entire pinion rotates counter-clockwise.

46. A
The equation of levers is
Load × Load distance = Force × Force distance

Here, the load is represented by the symbol P (force is F). On the other hand, Load distance is 2 units and force distance = 3 units.

The mechanical advantage of levers is
MA = Load/Force = (Force distance)/(Load distance)

47. D
The ratio of the load to the effort is known as Mechanical Advantage (MA). It shows how many times easier it would be to perform an action using the simple machine compared to not using it. Mechanical advantage has no unit.

48. A
A lateral view of the Wheel and Axle system is shown on the right. The pivot is at the common center of the two circles. Force is on the left side and Load is on the right one. Therefore, this is an example of first class lever (Force – Pivot – Load).

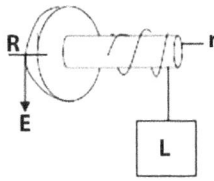
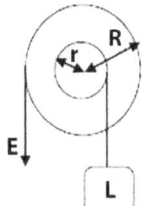

49. C
Sugar tongs have their turning point at one side and the load is on the other side. We must use a force between them to catch the sugar cubes. Therefore, this is an example of third class levers (Pivot – Force – Load).

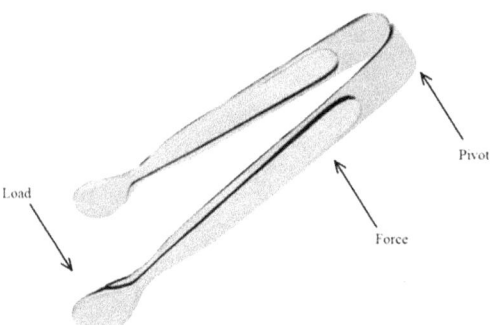

50. D
This is a combined system of simple machines composed of a Wheel and Axle and a Movable pulley system.

For the wheel and axle, the mechanical advantage is $MA_1 = R / r$. Here, MA1 = 3.

For movable pulleys, the mechanical advantage MA_2 is 2. So, the combined mechanical advantage is
$MA_{total} = MA_1 \times MA_2 = 3 \times 2 = 6$.

This means the force needed to lift the load is 6 time smaller than the load itself. So, F/W = 1/6

51. B
Since the system is in equilibrium, we have the total clockwise moment (turning effect) is equal to the counter clockwise one. The equation used in this situation is
$F_1 * d_1 = F_2 * d_2 + F_3 * d_3$

Or
$m_1 * g * d_1 = m_2 * g * d_2 + m_3 * g * d_3$
Simplifying gravity g from both sides and giving that d_1 = 3 units, d_2 = 1 unit and d_3 = 4 units, the equation becomes,
4 * 3 = 2 * 1 + m_3 * 4
12 = 2 + m_3 * 4
m_3 * 4 = 12 – 2 = 10
m_3 = 10/4 = 2.5 kg

52. C

First, we must find the weight of the hanging object on the left.

$W_{left} * d_{left} = W_{right} * d_{right}$
$W_{left} * 4 \text{ units} = 40N * 2 \text{ units}$
$W_{left} = (40N * 2 \text{ units}) / (4 \text{ units}) = 20N$

The total weight supported by the bar is 40N + 10N = 60N.

This value also corresponds to the reading of the dynamo-meter.

53. D

The bar weight tries to rotate the system clockwise. The weight is exerted at middle of the bar (d_w = 3 units). This is balanced by the pulling force of the rope exerted at 4 units away from the turning point (support). Thus, the upward force exerted by the rope is calculated by

$W_{bar} d_w = F_{rope} * d_F$

From the figure, you can see that the rope is connected to a combined pulley system (one fixed and one movable). Thus, given that only the movable pulley provides a gain in force (precisely double), we have for F_{rope} = 2 * 3N = 6N. So, we can write

$W_{bar} * 3 \text{ units} = 6N * 4 \text{ units}$
$W_{bar} = (6N * 4 \text{ units}) / (3 \text{ units}) = 8N$

54. B

The mechanical advantage of a lever (second class in the specific case) is calculated in two ways:

MA = W/F   or   MA = d_F/dW

The second equation is more suitable here, as we have enough information. So,

MA = 1m/20cm = 1m/0.2m = 5

55. C

The effort is represented here through the letter F (force). Thus, given that in pivoting systems in equilibrium

$W \cdot d_w = F \cdot d_F$

The equation becomes, after the substitutions,

400 N · 20cm = 100N · d_F
d_F = (400 N · 20cm) / 100N
= 80cm

56. D

The door handle has a turning effect when a force acts on it. Therefore, it is an example of a Lever (first class) as the turning point is between the force and load (they are on opposite sides of the door).

57. D

To create equilibrium, we must have:

$P \cdot d_P = F \cdot d_F$

If the required force distance d_F is denoted by x, the load distance d_P is 120 – x. Thus, we can write

100 · (120 - x) = 20 · x

Simplifying both sides by 20, the equation becomes,

5 · (120 - x) = x

5 · 120 – 5 · x = x

5 · 120 = x + 5x

600 = 6x

X = 600/6 = 100cm

Circuits and Electricity

1. The same battery and the three identical bulbs having equal resistance are used in three different combinations. Which setup will the battery last the longest?

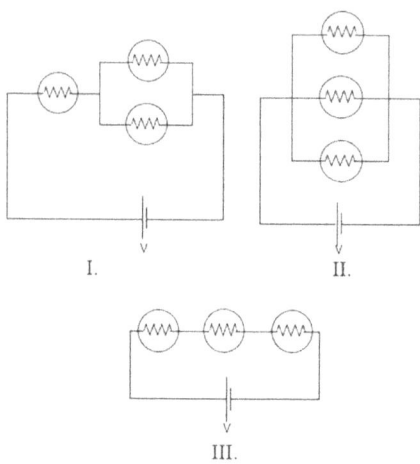

I.   II.

III.

a. In the first
b. In the second
c. In the third
d. All three are equal

2. What value does the voltmeter in the circuit read in volts if the circuit is closed?

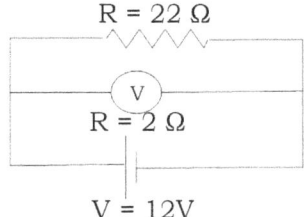

a. 13
b. 12
c. 11
d. 10

3. In a closed circuit

a. There is no electron flow
b. There is no current flow
c. Appliances do not work
d. Bulbs are glowing

4. A closed circuit is a circuit in which

a. The switch is turned on
b. The switch is turned off
c. There is no power source
d. There are no appliances connected

5. Which of the following explains how resistance of a conductor in a closed circuit is calculated?

    a. Ohm's Law
    b. Current Law
    c. Voltage Law
    d. Circuit Law

6. Which of the following is Ohm's Law equation for a closed circuit?

    a. $F = m \cdot a$
    b. $E = m \cdot c2$
    c. $x = v \cdot t$
    d. $R = V / I$

7. What three factors does resistance of a conductor depend?

    a. Density, length, color
    b. Color, weight, type of material
    c. Length, weight, color
    d. Length, thickness, type of material

8. The _____ of a material represents the resistance of a 1m long conductor which is 1 m² thick.

    a. Density
    b. Resistivity
    c. Resistance
    d. Voltage

9. **Which of the following statements about conductors is TRUE?**

   a. Human body is a good conductor of electricity
   b. Air is the worst conductor of electricity
   c. Metals are good conductors of electricity
   d. A very good conductor of electricity produces free electricity

10. **What is the result of increasing the temperature of a conductor?**

   a. An increase in the resistance of the conductor
   b. A decrease in the resistance of the conductor
   c. No change in the resistance of the conductor
   d. Total elimination of the conductor's resistance

11. **What is the opposition of a conductor to the flow of electrons called?**

   a. Density
   b. Resistance
   c. Voltage
   d. Current

12. What is the current in the circuit below?

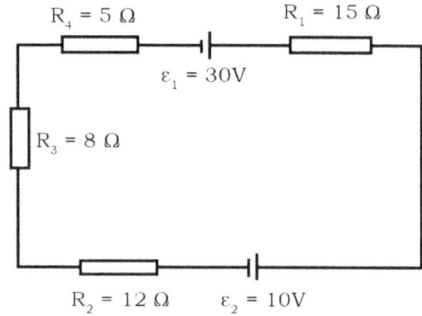

a. 0.5A
b. 1A
c. 2A
d. 4A

13. Which of the followings is equivalent to watt?

a. Volt/Ampere
b. Volt/Ohm
c. Volt x Ampere
d. Ohm x Ampere

14. An electric heater draws 10 A current from a 210 V source. How long will it take (in seconds) for the heater to heat 1 kg of water from 40°C to 50°C?
(C_{water} = 4200J/kg°C)

a. 20
b. 25
c. 200
d. 250

CIRCUITS AND ELECTRICITY

15. Which one is NOT the equation for Joule's Law?

a. $P \cdot t$
b. $I^2 \cdot R \cdot t$
c. $V \cdot I \cdot t$
d. $I \cdot R$

16. A 1000 Watt electric heater draws 5 A current. What is the resistance of the heater (in Ohms.)?

a. 25 Ω
b. 40 Ω
c. 200 Ω
d. 240 Ω

17. How many kilowatt-hours of energy does a 60-Watt light bulb consume in 2.5 hours?

a. 0.24
b. 0.15
c. 1.5
d. 2.4

18. Which of the following are effects of electricity?

I. Lighting
II. Physical
III. Magnetic
IV. Lifting

a. I,II,III
b. I,III
c. II,IV
d. II,III,IV

19. How many possible electrical paths are there in the circuit shown below?

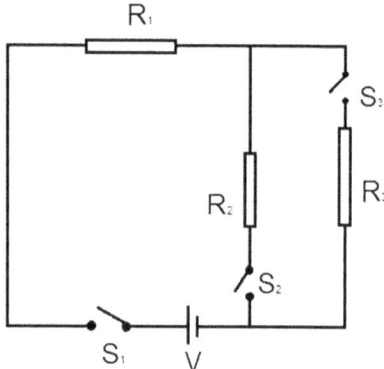

a. 1
b. 2
c. 3
d. 4

20. Which statement is CORRECT about the electric path in a circuit?

a. Electric path depends on the design of the circuit

b. Electricity prefers to choose the path in which it encounters more resistance

c. Electricity avoids passing in paths where there is no resistor

d. The current path through a circuit start from the negative terminal of the battery and ends to the positive one

21. Which is the easiest path for electricity to flow through the circuit below (when all switches are closed) if $R_1 = 2R_2 = 3R_3$?

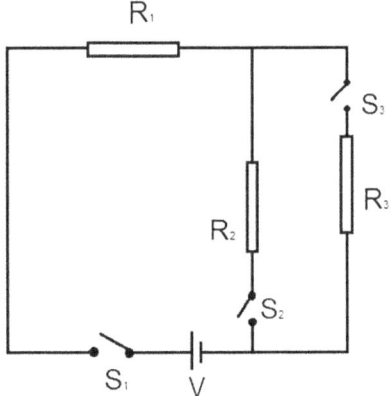

a. First through R_1, then through R_2
b. First through R_2 and then through R_3
c. First through R_3 and then through R_2
d. First through R_1, then through R_3

22. Which statement regarding the electric path is incorrect?

a. The sum of all currents entering a node in a circuit is equal to the sum of the currents leaving the node

b. Electric path shows the direction of the electrons flow around a circuit

c. When 3 resistors are connected in parallel, the current flows in three different paths

d. The resistance encountered by the current during its flow through a circuit depends on the length of the path

23. How many directions does the current flow through the circuit below with the current position?

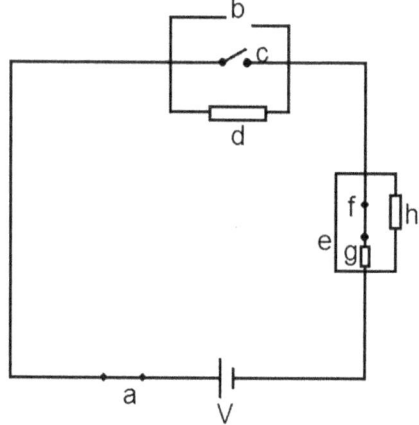

a. 1
b. 2
c. 3
d. 4

24. What is the current in the main branch of the circuit shown below if all resistors are 12Ω each and the resistance r of the battery is 2Ω? Take the potential difference of the source equal to 24 V.

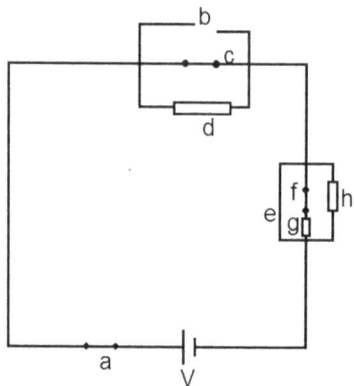

a. 12A
b. 4A
c. 3A
d. 2A

25. Which of the following statement about electric path is CORRECT?

a. The number of branches in an electrical circuit is always equal to the number of current paths

b. The number of branches in an electrical circuit is always greater than the number of current paths

c. The number of branches in an electrical circuit is always smaller than the number of current paths

d. The number of branches in an electrical circuit can be greater than or equal to the number of current paths but it cannot be smaller

26. Which one(s) of the lamps in the circuit aside will light, when the switches 1, 3 and 4 are closed?

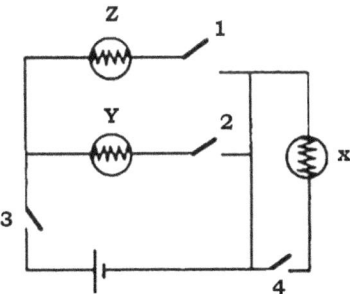

a. Z only
b. X and Y
c. X and Z
d. X only

27. If a switch is open

a. There is no current flowing through the circuit
b. There is some current flowing through the circuit
c. There is a lot of current flowing through the circuit
d. The circuit breaks due to overload

28. Which one(s) of the lamps in the circuit aside will glow, when the switches 1 and 4 are closed?

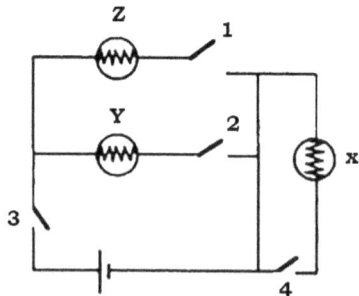

a. Z only
b. Z and Y
c. None of the lamps will glow
d. X only

29. Which of the statements below does not correctly describe a fuse?

a. It is used to change the value of current in a circuit
b. It has a high resistance
c. It is placed in the live wire
d. It has a low melting point

30. A fuse is a device used to

a. Change the voltage is a circuit
b. Protect electrical appliances from overload
c. Increase the power of an appliance
d. Produce electricity

31. The maximum current a fuse can hold is

a. Slightly lower than the operating current of the circuit
b. Much lower than the operating current of the circuit
c. Slightly higher than the operating current of the circuit
d. Much higher than the operating current of the circuit

32. Which circuit element shown below shows the symbol of fuse in the circuit below?

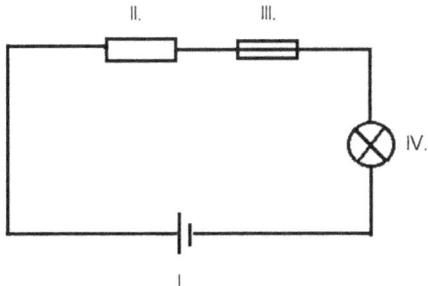

a. I
b. II
c. III
d. IV

33. A good fuse usually holds about 0.5A more current than the operating current of the circuit. What fuse must be used in the circuit shown below?

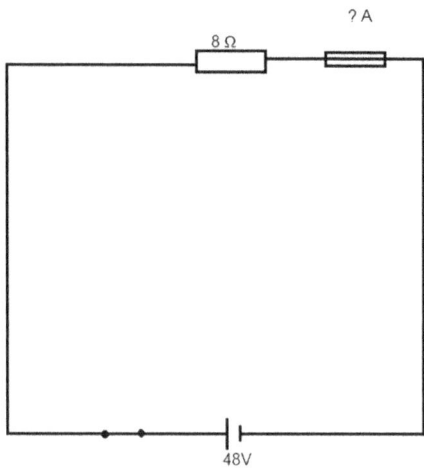

a. 5.5 A
b. 6.0 A
c. 6.5 A
d. 8.5 A

34. The thin wire inside a fuse has

a. High resistance and low melting point
b. High resistance and high melting point
c. Low resistance and low melting point
d. Low resistance and high melting point

35. Which statement below about fuse is correct?

   a. A fuse is a modern type of resistor
   b. A circuit breaker is a modern type of fuse
   c. A resistor is a modern type of fuse
   d. A fuse is a modern type of circuit breaker

36. Which statement is CORRECT about electric circuits?

   a. There is no electrons flow through an open circuit
   b. When the circuit is open, the bulbs glow brighter than when the circuit is closed
   c. It is called "open circuit" because there is an open (free) path for the electricity to flow through the circuit
   d. An open circuit exists when the switch is ON

37. When you turn a switch ON, you have ...

   a. Closed the way for the current to flow around the circuit
   b. Opened the way for the current to flow around the circuit
   c. Obtained an open circuit
   d. Options B and C

38. Which figure(s) below represent(s) an open circuit?

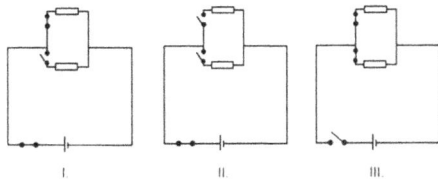

a. Only II
b. Only III
c. II and III
d. I and III

39. What is the minimum number of switches that must turn OFF to obtain an open circuit in the figure below?

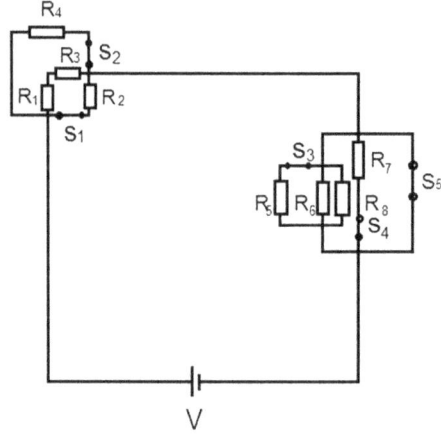

a. 2
b. 3
c. 4
d. 5

40. Which of the following actions represents the opening of a circuit?

a. Opening a door
b. Stopping a moving car
c. Turning off the lights
d. Both B and C

41. Which of the following examples represents an open circuit?

a. One of the resistors connected in series (in a series circuit) is defective and is not working

b. One of the resistors connected in parallel (in a parallel circuit) is defective and is not working

c. A short-circuit occurs due to a defective circuit element

d. Both A and C

42. The easiest way to obtain an open circuit is

a. To remove an element from the circuit

b. To remove all elements from the circuit

c. To remove the power source from the circuit

d. To turn off the switch

43. Which statement is true about the electric circuit shown below?

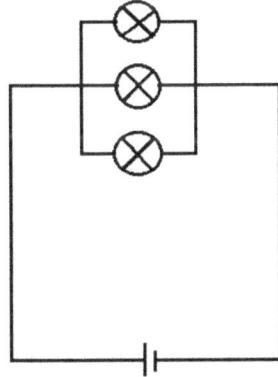

a. If one bulb breaks, the other two bulb will turn off

b. If one bulb breaks, the other bulbs will glow brighter

c. If one bulb breaks, the other bulbs will glow dimmer

d. Nothing will change in the other bulbs if one bulb breaks

44. The equivalent resistance in a parallel circuit is

   a. Greater than the resistance of each single resistor
   b. Smaller than the resistance of each single resistor
   c. Equal to the resistance of each single resistor if they are identical
   d. The sum of the resistances of all resistors in the circuit

45. What is the value of resistance R in ohms, if the equivalent resistance of the circuit is 1 Ω?

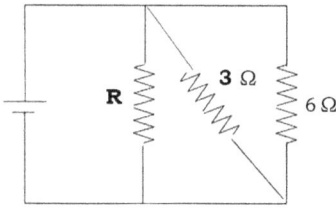

   a. 0
   b. 2
   c. 3
   d. 18

46. Find the ratio of the current in the second branch to the current in the third branch in figure; I_2/I_3=?

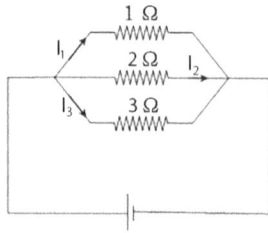

a. 3/2
b. 3/1
c. 2/3
d. 1/3

47. Which of the following statements is true?

a. Voltmeter has a low resistance and it is connected in series in the circuit

b. Voltmeter has a high resistance and it is connected in series in the circuit

c. Voltmeter has a low resistance and it is connected in parallel in the circuit

d. Voltmeter has a high resistance and it is connected in parallel in the circuit

48. Which statement is false regarding the circuit elements connected in parallel?

a. All circuit elements connected in parallel have the same potential difference

b. If one of the circuit elements connected in parallel breaks, the other elements operate regularly

c. We can connect in parallel circuit elements with different operating power

d. We cannot connect in parallel circuit elements with different operating power

Knife Switches

49. What position must knife switches be installed?

    a. In such a way so gravity closes them easily
    b. In such a way so gravity won't tend to close them
    c. In such a way so they remain always open
    d. In such a way so they remain always closed

50. Knife switches are mostly used for

    a. Isolation switches
    b. Disconnecting switches
    c. Connecting switches
    d. Conduction switches

51. How must the blades of a knife switch be connected?

    a. The position of the blades is not important
    b. When open, the blades will be "dead"
    c. Only vertically
    d. Only horizontally

52. What type of switches are knife switches?

    a. Mechanical
    b. Electronic
    c. Rotary
    d. Digital

53. Which category does the knife switch shown below belong?

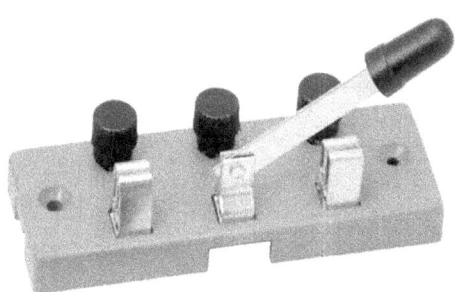

a. SPST (single pole Single throw)
b. SPDT (single pole double throw)
c. DPST (double pole, single throw)
d. DPDT (double pole double throw)

54. Which type of switch is the knife-switch shown below?

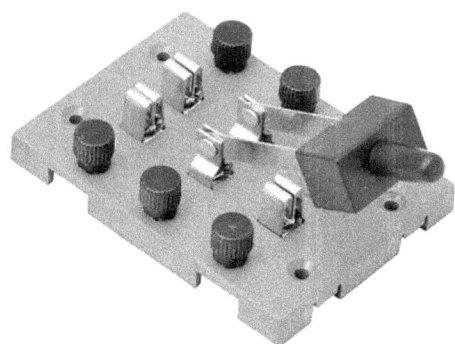

a. SPST (Single Pole Single throw)
b. SPDT (single pole double throw)
c. DPST (double pole, single throw)
d. DPDT (double pole double throw)

55. Below is a circuit diagram symbol of a knife switch. What type of switch is it?

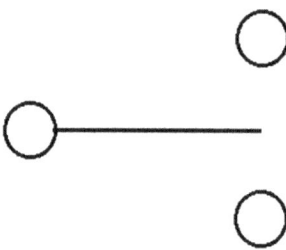

a. SPST (Single Pole Single throw)
b. SPDT (single pole double throw)
c. DPST (double pole, single throw)
d. DPDT (double pole double throw)

Positive and Negative Poles

56. The positive pole of a battery has a surplus of _____

a. Protons
b. Electrons
c. Neutrons
d. Atoms

57. Electric current flows from _____ to _____ pole of an electric source

    a. Positive to negative
    b. Negative to positive
    c. Negative to neutral
    d. Positive to neutral

58. The direction of electrons flow through a circuit is from _____ to _____ terminal of the battery.

    a. Positive to negative
    b. Negative to positive
    c. Negative to neutral
    d. Positive to neutral

59. Which letter in the circuit elements shown below represents the negative pole of a cell?

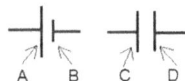

    a. A
    b. B
    c. C
    d. D

60. Negative pole of a battery is the one with _____ electric potential.

    a. The highest
    b. The lowest
    c. Zero
    d. Infinity

61. If two cells are connected in series, the _____ terminal of one cell is connected to the _____ terminal of the other.

    a. Positive, negative
    b. Positive, positive
    c. Negative, negative
    d. Neutral, negative

62. Which statement about the poles of a battery is correct?

    a. Poles of a battery are always in the opposite direction of the battery
    b. A battery may contain only one pole
    c. Poles of a battery must be connected in the shortest possible path
    d. If you connect the poles of a battery by a conducting wire, the battery is discharged very fast

Series Circuit

63. What is the potential difference across the resistor R_3 in the circuit in figure?

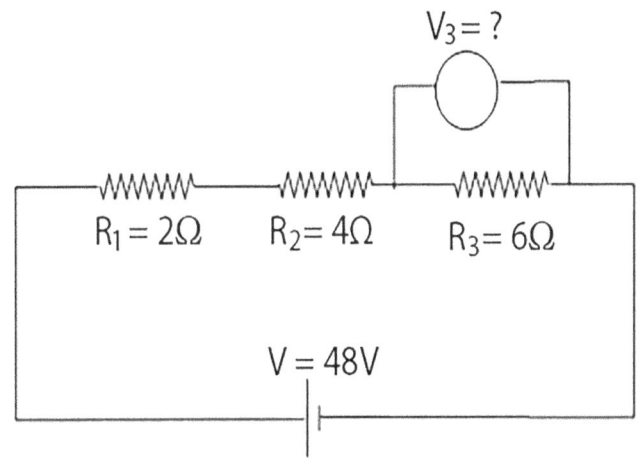

a. 24V
b. 12V
c. 8V
d. 4V

64. What is the relation between i_1, i_2 and i_3 according to the figure aside?

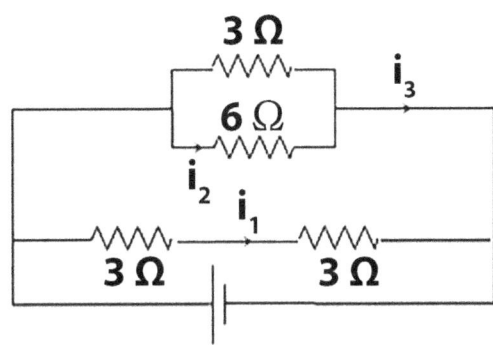

a. $i_1 > i_2 > i_3$
b. $i_1 > i_3 > i_2$
c. $i_2 = i_1 > i_3$
d. $i_3 > i_1 = i_2$

65. What is the ratio of the equivalent resistance for 3 identical resistors connected in series to the case when they are connected in parallel?

a. 1/9
b. 1/3
c. 3
d. 9

66. The advantage of a series circuits compared to a parallel circuit is that series circuits make the circuit elements ...

a. Operate at higher potential difference
b. Operate at higher current
c. Operate at higher power
d. Operate independently from each other

67. A shortcoming of series circuits is that

a. If one circuit element fails to operate, the other elements operate at lower current
b. If one circuit element fails to operate, the other elements operate at higher current
c. If one circuit element fails to operate, the other elements cannot operate
d. If one circuit element fails to operate, the other elements operate at lower potential difference

68. Which statement below is correct about short circuits?

   a. A short circuit makes the value of current very small

   b. A short circuit makes the value of current very large

   c. A short circuit makes the value of current equal to infinity as there is no resistance to oppose the movement of electric charges

   d. A short circuit makes the value of current zero

69. A short circuit causes harm to the electric devices connected in the circuit because of

   a. Excessive current generated in the circuit

   b. Excessive potential difference generated in the circuit

   c. Excessive force exerted in the operating devices connected in the circuit

   d. Excessive resistance generated in the circuit

70. What is the operating current in the main branch of the circuit shown below? All resistors are identical and their magnitude is equal to 4 ohms.

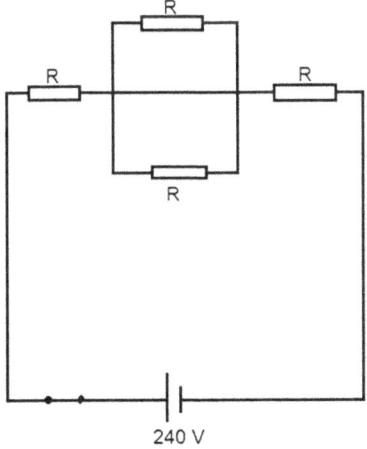

a. 24 A
b. 30 A
c. 60 A
d. 120 A

71. What is the ratio of the equivalent resistance in the circuit shown below when the switch B is closed, to the case when the switch is open? All resistors are identical.

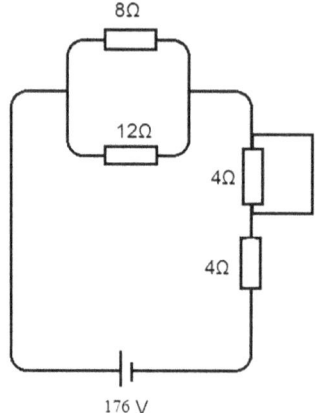

a. 11/6
b. 4/3
c. 8/11
d. 11/6

72. The total resistance in a short circuit is

a. Zero
b. Very small
c. Equal to the resistance of the operating device
d. Very large

73. A "short circuit" is so called because the current path

   a. Is the shortest possible
   b. Is the easiest possible
   c. Is not determined
   d. Is interrupted

74. What is the value of the current in the main branch of the circuit shown below?

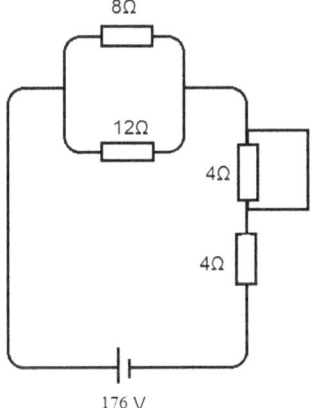

   a. 20 A
   b. 13.75 A
   c. 6.28 A
   d. 6.29 A

75. A toggle switch is a type of _____ switch

   a. Electronic
   b. Mechanical
   c. Digital
   d. Automatic

76. How many positions does a toggle switch have?

 a. 2
 b. 3
 c. 4
 d. Many

77. A toggle switch is activated through a

 a. Projecting lever
 b. Laser
 c. Remote control
 d. Digital button

78. Which of the switches shown is NOT a toggle switch?

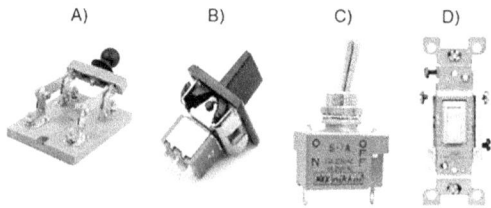

 a. Switch A
 b. Switch B
 c. Switch C
 d. Switch D

79. Which of the following actions is performed using a toggle switch?

a. Dimming the light in a room

b. Moving the lights around in a pub

c. Turning on or off a bulb

d. Moving the cursor through the mouse on the monitor of computer

80. Which of the following statements regarding toggle switch is correct?

a. A toggle switch is easy to use

b. A toggle switch has many positions available

c. A toggle switch is an expensive device as it contains many mechanisms

d. Only one toggle switch can be installed in an electric circuit

Answer Key

1. B
The battery will last longer if the total resistance in the circuit is the smallest. This occurs when all resistors are connected in parallel.

2. B
The potential difference in a parallel combination of circuit elements (as shown) is the same as the potential difference of the source.

3. D
A closed circuit is a circuit in which the current is flowing across it. Therefore, all appliances (including) bulbs operate in a closed circuit.

4. A
A circuit is closed when the switch is turned on.

5. A
In a closed circuit, resistance is calculated using Ohm's Law: $R = V / I$ where V is the potential difference (voltage) and I is the electric current.

6. D
In a closed circuit, the resistance is calculated by the Ohm's Law: $R = V / I$ where V is the potential difference (voltage) and I is the electric current.

7. D
The equation of the resistance of a conductor is
$R = \rho \cdot L/A$

Where ρ is the resistivity (a property of material), L is the conductor's length and A is the conductor's cross-sectional area.
This means that density of a conductor depends on 3 factors: conductor length, thickness, and type of material.

8. B
Resistivity ρ is a property of material that shows the resistance provided by a conductor per unit length and unit cross

sectional area. It is given in specific tables (like density, specific heat capacity, etc.).

9. C
Choice A is incorrect. Despite having some conductive ability, the human body is a poor conductor of electricity.
Choice B is incorrect. Vacuum is the worst conductor of electricity, not air.
Choice C is correct. Metals are good conductors of electricity.
Choice D is incorrect. A very good conductor of electricity allows electricity flow easily through it - it does not produce electricity by itself.

10. A
An increase in temperature of the conductor brings a more rapid vibration of its atoms. Therefore, it becomes more difficult for electricity (free electrons) to flow through the conductor. As a result, the resistance increases.

11. B
By definition, The opposition of a conductor to the flow of electrons is called resistance. It is measured in Ohms (Ω).

12. A
Taking the clockwise direction as positive and applying the Kirchhoff Law of branches, the equation becomes,
$\varepsilon_1 - \varepsilon_2 = I \cdot (R_1 + R_2 + R_3 + R_4)$

Substituting the values, we get

$30 - 10 = I \cdot (15 + 12 + 8 + 5)$
$20 = I \cdot 40$
$I = 0.5A$

13. C
Watt is the unit of power. Since power is calculated by multiplying potential difference and current, the unit of power is Volt × Ampere. It is the unit used to measure the load of appliances, which consume electricity.

14. A
From the law of energy conservation, we have:
Heat released by the power source = Heat gained by the water.

Thus, we have
$V \cdot I \cdot t = m \cdot c \cdot \Delta T$

Substituting the values, the equation becomes,

$210 \cdot 10 \cdot t = 1 \cdot 4200 \cdot (50 - 40)$
$2100 \cdot t = 42000$
$T = 42000/2100$
$= 20s$

15. D
Joules law (regarding the electric energy E) is given by the equation
$E = P \cdot t$
$= V \cdot I \cdot t$
$= I \cdot R \cdot I \cdot t$
$= I^2 \cdot R \cdot t$

16. B
First, we find the potential difference of the circuit. Thus,
$V = P/I = 1000W/5A = 200V$

So, the resistance (using the Ohm's law) is
$R = V/I$
$= 200V/5A$
$= 40\Omega$

17. B
From Joule's law, we have
Energy = Power · Time

We must convert watts in kilowatts. Thus, 60 W = 0.06 kW.
So,
Energy = 0.06 kW · 2.5h = 0.15 kW · h

18. D
"Physical" is not an effect of electricity. The other 3 are related to electricity. The lighting effect of electricity is observed when you turn on a lamp. Magnetism is an effect of electricity observed in electromagnets. Lifting is an effect of electricity observed in cranes.

19. B
The possible paths are observed when all switches are closed. Thus, the first path is

Power source – S_1 – R_1 – R_2 – S_2 - Power source

The second path is

Power source – S_1 – R_1 – S_3 – R_3 - Power source

There are two possible paths in this circuit.

20. A
Choice A is correct. Electric path depends on the design of the circuit.
Choice B is incorrect. Electricity chooses the path of least resistance.
Choice C is incorrect. Electricity flows easily in paths where there is no resistor. Here, a short circuit occurs.
Choice D is incorrect. The current path through a circuit starts at the positive terminal and ends at the negative one. Do not confuse the direction of current with the direction of electrons flow, which are opposites.

21. D
Electricity chooses the easiest path to flow, i.e. the path which offers less resistance. So, the easiest path of electricity (where most of the current flows) is
V – S_1 – R_1 – S_3 – R_3 – V

22. B
Choice A is correct. This is the Kirchhoff law of nodes.
Choice B is incorrect. Electric path shows the direction of electric current which is opposite to that of electrons flow in a circuit.
Choice C is correct. When 3 resistors are connected in parallel, the current flows in three different paths.
Choice D is correct. Longer the path, higher the resistance provided by the conductor to the current flow.

23. A
The point b represents a broken wire. No electricity can flow here.

The point c represents an open switch. Electricity cannot flow here either.

Thus, electricity after passing through the resistor d, flows through the point e (as it represents a short circuit) avoiding thus the points f, g and h.

So, there is only one possible direction of current flow through the actual circuit.

24. A
There is a short circuit where current passes through the points c and e almost without resistance (if the resistance of wire is not considered). Thus, the only resistance is provided by the source.

Thus,
$I = V/r = 24V/2\Omega = 12A$

25. D
Choice A is incorrect. It is enough to provide a counter-example to reject a claim. Thus, for the first statement we can take the case when a switch is open. Although there is a branch in that place, no current flows.
Choice B is incorrect. If there is only one resistor in a simple circuit (and therefore a single branch) there is also one path for the electricity flow.
Choice C is incorrect. The number of branches in an electrical circuit cannot be smaller than the number of current paths.
Choice D is correct. The number of branches in an electrical circuit can be greater than or equal to the number of current paths but it cannot be smaller.

26. A
The current from the power source will flow through switch 3, then through the lamp Z, then from the switch 1, then through the short circuit (without passing through the lamp X and switch 4) and ending up to the power source again. So, only the lamp Z will glow.

27. A
An open switch means no current flowing through the circuit (not a circuit break due to overload).

28. C
If the switch 3 is not closed, no current will flow through the circuit. Therefore, none of the lamps will glow.

29. A
A fuse is a circuit element used to protect electric appliances from excessive current. It has a high resistance and a low melting point, so that it can easily break when the current exceeds the normal value. Fuse is placed in the live wire. It cannot change the value of current in the circuit.

30. B
A fuse is a circuit element used to protect electric appliances from the excessive current. It has a high resistance and a low melting point, so that it can easily break when the current exceeds the normal value. Fuse is placed in the live wire. It cannot change the value of current or voltage in the circuit, neither can produce electricity by itself or increase the power in the appliance.

31. C
The maximum current in a fuse is slightly higher than the operating current of the appliance it aims to protect. This is because if the current becomes too large, the fuse breaks and the appliance will save from defects.

32. C
I is the symbol of cell (power source).
II is the symbol of resistor.
III is the symbol of fuse.
IV is the symbol of bulb.

33. C
The operating current of the circuit (using the Ohm's law) is
$I = V/R = 48V/8\Omega = 6A$
Thus, a 6A+ 0.5A = 6.5A fuse must be installed in the circuit.

34. A

A fuse is a circuit element used to protect electric appliances from the excessive current. It has a high resistance and a low melting point, so that it can easily break when the current exceeds the normal value.

35. B

Choice A is incorrect. A fuse is not a resistor; it is a protective device from excessive current.
Choice B is correct. A circuit breaker is a modern type of fuse, which switches off when there is an issue in the circuit.
Choice C is incorrect. A resistor is not a fuse but a circuit element, which is used to convert electricity into heat.
Choice D is incorrect. A fuse is old fashioned compared to circuit breaker.

36. A

Choice A is correct. When a circuit is open, the switch is off. Therefore, there is no electron flow through an open circuit
Choice B is incorrect. When the circuit is open, the bulbs are off.
Choice C is incorrect. It is called an "open circuit" because there is an open gap which prevents the current flow through the circuit.
Choice D is incorrect. An open circuit exists only when the switch is OFF.

37. B

Choice A is incorrect. When you turn a switch ON, you allow the electricity flow through the circuit. Therefore, you open the way to the current flow around. This means B is correct.
Choice C is incorrect. When the switch is ON, there is a closed circuit.

38. C

An open circuit is a circuit in which electricity is unable to flow. In the first figure, electricity can flow through the upper resistor in the parallel branch, while in the other two figures electricity cannot flow through the circuit. In the second figure current cannot pass through the parallel branch and in the third figure current cannot flow through the open switch near the source.

39. B
Switches S_1 and S_2 are not relevant as the current can flow through R_1 and R_3 regardless the position of these two switches. Thus, to obtain an open circuit, i.e. to block the flow of current, you must turn off all the three switches in the right.

40. C
Opening a door is not an example related to the flow of current in an electric circuit.
Choice B is incorrect. You can stop a car without turning it off (at traffic lights for example). So, the car's electric circuit is not turned off to obtain an open circuit.
Choice C is correct. Turning off the lights is an example of an open circuit.

41. A
Choice A is correct. When one resistor connected in series is defective, the electricity flow stops as there is a single path for electricity in series circuits. Therefore, the equation becomes an open circuit.
Choice B is incorrect. A parallel circuit provides more than one path for the electricity to flow. This means when a resistor is faulty, the current flows through the other paths and the circuit remains closed.
Choice C is incorrect. In a short circuit, electricity flows easily and in large amounts through the circuit without encountering resistance. This means the circuit is closed.

42. D
Choice A is incorrect. You can remove an element but the circuit may contain other elements that keep the circuit closed.
Choice B is incorrect. A circuit with all elements removed is useless.
Choice C is incorrect. If you remove the power source from a circuit, it is not an electric circuit.
Choice D is correct. The purpose of switches is to obtain open or closed circuits easily, through a simple action.

43. B
This is an example of parallel circuit, in which when one element is faulty, the other elements continue operating at

higher capacity because the current intended for the faulty branch distributes in the remaining branches.
In the specific case, when a bulb is faulty, the other bulbs glow brighter.

44. B
The equation of the equivalent resistance in a parallel circuit is
$1/R_{eq} = 1/R_1 + 1/R_2$
For simplicity, we can consider two identical resistors, R. Thus, we have
$1/R_{eq} = 1/R + 1/R = 2/R$
So,
$R_{eq} = R/2$

45. C
This is an example of a parallel circuit as the current splits in three branches at the upper node. From the equation of parallel circuits, we have
$1/R_{eq} = 1/R_1 + 1/R_2 + 1/R_3$

Substituting the values, the equation becomes,
$1/1 = 1/R + 1/3 + 1/6$
$1/R = 1/1 - 1/3 - 1/6$
$= 6/6 - 2/6 - 1/6$
$= 2/6$
Thus, $R = 6/2 = 3 \, \Omega$.

46. A
The current in a parallel circuit divides into branches, while the potential difference for all branches is equal. This means current and resistance are inversely proportional. So, in the upper branch, the current will be V/1, in the middle branch V/2 and in the lower branch, the current is V/3. So, the ratio of currents I2 and I3 is
$I_2 / I_3 = (V/2) / (V/3) = 3/2$

47. D
A voltmeter is a device used for measuring the potential difference across a circuit element. It must have a very high resistance to allow only a very small amount of current to flow through it and therefore, to not affect the value of current

flowing through the circuit. A voltmeter must be connected in parallel to the circuit element, i.e. one wire at the beginning and the other wire at the end, to create a bypass for the current as shown.

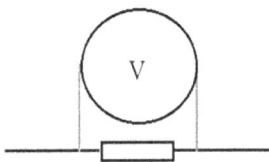

48. D
From theory, it is known that all circuit elements connected in parallel have the same potential difference, while the current splits in different branches. Thus, choice A is correct.

Choice B is correct as since current is divided into branches, If one of the circuit elements connected in parallel breaks, the other elements operate regularly.

Choice C is correct. We can connect various electric appliances with different power rating either in series or in parallel. No limitation exists on this.

Knife Switches

49. B
A knife switch must be installed in such a way so gravity won't tend to close it. This prevents any unexpected interruption of the current flow in the circuit.

50. A
An isolation switch used to isolate an electric circuit from its source of power. It has no interrupting rating and is intended to be operated only after the circuit has been opened by some other means. Knife switches are used for this purpose.

51. B
The blades of a knife switch must be connected in such a way so that when open, the blades will be "dead." This will not effect the circuit.

52. A
Knife switches are mechanical switches as they are moved by hand.

53. B
Since the knife switch shown has only one blade, it is a single pole, with two closing positions. This means it is a double-throw model.

54. D
Since the knife switch shown has two blades, it is a double pole, and has two closing positions. This means it is a double throw model.

Therefore, the knife switch shown is DPDT (double pole double throw).

55. B
The straight line is the symbol of blade. This means the knife switch is a single pole, as it has only one blade.

It is obvious that the switch can be closed in two positions (actually it is opened). So, the knife switch is a double throw model.
Therefore, the above circuit symbol represents a SPDT (single pole double throw) knife switch.

Positive and Negative Poles

56. A
Protons are regarded as positive electric charges and electrons as negative ones.

Therefore, the symbol (+) in the battery shows that there is a surplus of protons in that region of the battery.

57. A
The direction of electric current flow is from the (+) to (-) terminal of the battery. Therefore, electric current flows from positive to the negative pole of an electric source.

58. B
The direction of electrons flow is from the places where there are more electrons, to the places where there are fewer electrons. It is opposite to the direction of current flow which is from plus to minus for historical reasons.

59. B
The first circuit symbol (a long and a short line) represent a cell, where the longest line represents the positive terminal and the shortest one, the negative poles.

The second symbol of circuit element (the one with two equal parallel lines) represents a capacitor.

60. A
Electrons flow from higher potential to lower potential. As electrons flow from the negative terminal of the battery to the positive, the negative pole of a battery represents the highest electric potential.

61. A
When two cells are connected in series, the positive terminal of one cell is connected to the negative cell of the other, because the same current must flow through the circuit.

This setup increases the potential difference of the circuit as $V_{total} = V_1 + V_2 + ...$

62. D
Choice A is incorrect. Some batteries have their poles at the same side.

Choice B is incorrect. All batteries have two poles: positive and negative.

Choice C is incorrect. There is no rule the poles of battery must be connected in the shortest possible path.

Choice D is correct. If you connect the poles of a battery by a conducting wire, the battery is discharged very fast as a short circuit is created. It is known that the current in a short circuit is very high, which brings in a fast discharge of the battery.

Series Circuits

63. A

First, calculate the equivalent resistance of the series circuit shown. Thus,

$R_{eq} = R_1 + R_2 + R_3$

$2Ω + 4Ω + 6Ω$

$= 12Ω$

Second, we need to calculate the current flowing in the circuit.

$I = V/R_{eq} = 48V / 12Ω = 4A$

Now, we calculate the potential difference across the 6 ohms resistor, which is the value shown by the voltmeter.

64. D

The equivalent resistance in the parallel circuit is

$1/R_p = 1/3 + 1/6 = (2 + 1)/6 = 3/6$

Thus, $R_p = 6/3 = 2Ω$.

The current in the parallel branch is distributed according the inverse proportionality rule, i.e. in the 3Ω resistor will flow double the current flowing in the 6Ω resistor.

The same rule is applied to determine the current flowing through the two resistors connected in series and the parallel branch (which forms a big parallel branch). Thus, in the series branch, we have
$R_s = 3Ω + 3Ω = 6Ω$.

The equations become:

$i_1 = 1/3\ i_3$ (as $6Ω = 3 * 2Ω$)

$i_3 = (2 + 1)\ i_2 = 3i_2$

Rearranging, the equation becomes,

$i_3 = 3i_1$ and $i_3 = 3i_2$

So, we have $i_3 > i_1 = i_2$

65. D

If we have three identical resistors connected in series, we have

$R_s = R + R + R = 3R$

When the same resistors are connected in parallel, we have for the equivalent resistance:

$1/R_p = 1/R + 1/R + 1/R = 3/R$

Thus, $R_p = R/3$. The ratio between R_s and R_p is

$R_s / R_p = 3R / (R/3) = 9$

66. B

A series circuit has the same current in all components. This makes the current flow as a whole, without splitting into branches. This does not result in a higher potential difference and power, as each element has its own potential difference whose sum is the total potential difference of the power source.

67. C

Despite the advantages it offers, a series circuit has its shortcomings, the most important of which is that if one circuit elements fails, the other elements cannot operate. This is because all circuit elements in a series circuit have the same current and when one element is defective, the current flow stops.

68. B

When a short circuit occurs, the resistance of the circuit element is bypassed by the current, which chooses the easiest path to flow. There is only a very small resistance in the circuit provided by the conducting wire. From the Ohm's Law, we know that

$I = V/R$

Where V is constant; it is provided by the power source and does not depend on whether there is a short circuit or not. Therefore, the current increases to high values (but not to infinity), making the circuit very dangerous for the other devices installed in the circuit.

69. A

When a short circuit occurs, the resistance of the circuit element is bypassed by the current, which chooses the easiest path to flow. There is only a very small resistance in the circuit provided by the conducting wire. From the Ohm's Law, we know that

$I = V/R$

Where V is constant; it is provided by the power source and does not depend on whether there is a short circuit or not. Therefore, the current increases to high values (but not to infinity), making the circuit very dangerous for the other devices installed in the circuit.

70. A

Here, there is no parallel branch as the current flows directly through the free wire in the middle, overlooking the two parallel resistors (short circuit). So, we have only 2 resistors connected in series. Their equivalent resistance is

$R_{eq} = 4\Omega + 4\Omega = 8\Omega.$

So, the current in the main branch is

$I = V/R_{eq} = 240V / 8\Omega = 30A$

71. C

When the switch B is closed, a short circuit occurs in the upper branch. Thus, the only resistance is that of the branch on the right. Its value is

$1/R_p = 1/R + 1/R + 1/R = 3/R$

Thus, the equivalent resistance of the entire circuit is

$R_{eq}1 = R/3 + R = 4R/3$

When the circuit B is open, there is an additional parallel branch with two resistor to be included in the calculations. Using the same procedure as before, we find that its value is R/2. Thus, we have for the equivalent resistance of the whole circuit:

$R_{eq2} = R/3 + R + R/2$

$= (2R + 6R + 3R) / 6$

$= 11R / 6$

CIRCUITS AND ELECTRICITY

So, the ratio between the equivalent resistances in the two-above mentioned cases is

R_{eq1} / R_{eq2} = (4R/3) / (11R/6) = (6 · 4R) / (3 · 11R) = 24/33
= 8/11

72. B
When a short circuit occurs, the resistance of the circuit element is bypassed by the current, which chooses the easiest path to flow. There is only a very small resistance in the circuit provided by the conducting wire.

73. B
The current in short circuit chooses the easiest path to flow, skipping the appliances and flowing through the free wire, which offers only a very small resistance due to its internal properties.

Only in few cases, the term "short" refers to the path length; in most cases it is related to the ease of flow.

74. A
The 4Ω resistor is not counted as a short circuit occurs in that place. Thus, we have only a parallel branch and another single resistor to consider.

For the parallel branch we have:

$1/R_p$ = 1/8 + 1/12 = (3 + 2) / 24 = 5/24

So,

R_p = 24/5 Ω = 4.8Ω

The total equivalent resistance of the circuit is

R_{eq} = 4Ω + 4.8Ω = 8.8 Ω

The current in the main branch is calculated using the Ohm's Law:

I = V/R_{eq} = 176V / 8.8Ω = 20A

75. B
A toggle switch is a mechanical switch with only 2 positions: ON and OFF. It is a user-friendly switch operated by hand. Look at the figure:

76. A

A toggle switch is a mechanical switch with only 2 positions: ON and OFF. It is a user-friendly switch operated by hand. Look at the figure:

77. A

A toggle switch is an electric switch operated by a lever that is moved up and down. It has only two possible positions: ON and OFF.

78. A

The switch A is not a toggle switch. It is a knife switch of DPST (double pole single throw) type as it has two blades but only one closed position.

All the other switches are toggle switches with only two available positions: ON and OFF.

79. C

Dimming the light in a room includes a rotational motion of the switch which includes many possible positions. The same thing is true for moving lights in a pub. This is made through complex mechanisms that involve many positions. Choices A and B are incorrect.

Choice C is correct because a toggle switch is a mechanical switch with only two positions: ON or OFF.

Choice D is incorrect. No switching mechanism is involved in moving the cursor through the mouse on the monitor of computer.

80. A

Choice A is correct. A toggle switch is easy to use as it includes only two position: ON and OFF.

Choice B is incorrect. A toggle switch is suitable but not because it has many positions available (it has only two) but for other reasons.

Choice C is incorrect. A toggle switch is cheap and it includes only a few mechanisms.

Choice D is incorrect. We can install multiple toggle switches in the same circuit depending on the demands.

Basic Physics

	A	B	C	D	E		A	B	C	D	E
1	○	○	○	○	○	21	○	○	○	○	○
2	○	○	○	○	○	22	○	○	○	○	○
3	○	○	○	○	○	23	○	○	○	○	○
4	○	○	○	○	○	24	○	○	○	○	○
5	○	○	○	○	○	25	○	○	○	○	○
6	○	○	○	○	○	26	○	○	○	○	○
7	○	○	○	○	○	27	○	○	○	○	○
8	○	○	○	○	○	28	○	○	○	○	○
9	○	○	○	○	○	29	○	○	○	○	○
10	○	○	○	○	○	30	○	○	○	○	○
11	○	○	○	○	○						
12	○	○	○	○	○						
13	○	○	○	○	○						
14	○	○	○	○	○						
15	○	○	○	○	○						
16	○	○	○	○	○						
17	○	○	○	○	○						
18	○	○	○	○	○						
19	○	○	○	○	○						
20	○	○	○	○	○						

1. In which case shown below, is the work done against gravity is the greatest?

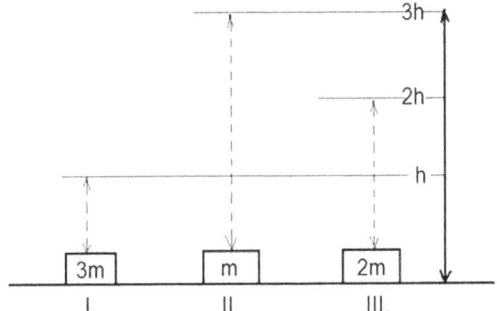

a. At I
b. At II
c. At III
d. At II and III

2. A 20 N horizontal force is used to push a 5kg object at 10 m away as shown. What is the work done by gravity on the object?

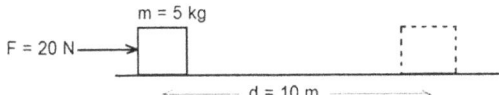

a. 200 J
b. 500 J
c. 50 J
d. 0

3. In which case below is the work done moving an object against gravity?

a. When moving the object downwards
b. When moving the object horizontally
c. When moving an object upwards
d. When an object is at rest

4. A crane lifts a 400 kg object at a height of 15 m above the original position. What is the work against gravity done by the crane? Take g = 10 N/kg.

a. 4,000 J
b. 6,000 J
c. 40,000 J
d. 60,000 J

5. If there is no friction, the minimum force F needed to move an object at a distance d along a horizontal plane is

a. Zero
b. Slightly greater than zero
c. Equal to the weight of the object
d. Slightly greater than the weight of the object

6. What is the work done against gravity to lift a 100 N heavy object up to the stairs shown?

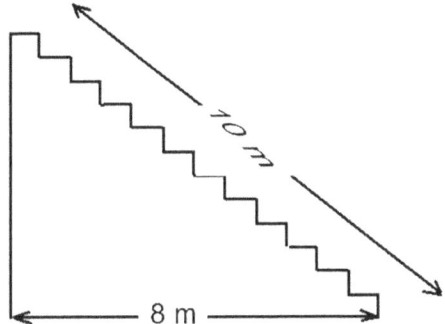

   a. 600 J
   b. 800 J
   c. 1000 J
   d. 1400 J

7. Which statement below is incorrect?

   a. The work done against gravity when throwing an object downwards is equal to the change in the gravitational potential energy of the object
   b. The work done against gravity when throwing an object upwards is equal to the change in the gravitational potential energy of the object
   c. The work done against gravity when moving an object horizontally is equal to the change in the gravitational potential energy of the object
   d. The work done against gravity when pushing a heavy object without being able to move it, is smaller than the change in the gravitational potential energy of the object

8. Newton's laws of motion consist of three physical laws that form the basis for classical mechanics. Which of the following is/are not included in these laws?

   a. Unless acted on by a force, a body at rest stays at rest.

   b. Unless acted on by a force, a body in motion will change direction and gradually slow until it eventually stops.

   c. To every action, there is an equal and opposite reaction.

   d. A body acted on by a force will accelerate in the same direction as the force at a magnitude that is directly proportional to the force.

9. A car starts from a full top and in 20 seconds is traveling 10/m per second. What is the acceleration?

   a. 0.5 m/sec^2

   b. 0.24 m/sec^2

   c. 1 m/sec^2

   d. 1.5 m/sec^2

10. If the space station travels 1000 meters in 5 seconds, how fast is it traveling?

   a. 100 meters/second

   b. 200 meters/second

   c. 50 meters/second

   d. 500 meters/second

11. How much force is needed to accelerate a car weighing 2,000 kg, at a rate of 3 m/s²?

   a. 6000 N
   b. 3000 N
   c. 2000 N
   d. 1000 N

12. Electricity is a general term encompassing a variety of phenomena resulting from the presence and flow of electric charge. Which of the following statements about electricity is/are true?

   a. Electrically charged matter is influenced by, and produces, electromagnetic fields.

   b. Electric current is a movement or flow of electrically charged particles.

   c. Electric potential is a fundamental interaction between the magnetic field and the presence and motion of an electric charge.

   d. An influence produced by an electric charge on other charges in its vicinity is an electric field.

13. Which of the following is/are not included in Ohm's Law?

   a. Ohm's Law defines the relationships between (P) power, (E) voltage, (I) current, and (R) resistance.

   b. One ohm is the resistance value through which one volt will maintain a current of one ampere.

   c. Using Ohm's Law, voltage is determined using V = IR, with I equaling current and R equaling resistance.

   d. An ohm (Ω) is a unit of electrical voltage.

14. The property of a conductor that restricts its internal flow of electrons is:

   a. Friction
   b. Power
   c. Current
   d. Resistance

15. In physics, _____ is the force that opposes the relative motion of two bodies in contact.

   a. Resistance
   b. Abrasiveness
   c. Friction
   d. Antagonism

16. What is the difference, of any, between kinetic energy and potential energy?

   a. Kinetic energy is the energy of a body that results from heat while potential energy is the energy possessed by an object that is chilled

   b. Kinetic energy is the energy of a body that results from motion while potential energy is the energy possessed by an object by virtue of its position or state, e.g., as in a compressed spring.

   c. There is no difference between kinetic and potential energy; all energy is the same.

   d. Potential energy is the energy of a body that results from motion while kinetic energy is the energy possessed by an object by virtue of its position or state, e.g., as in a compressed spring.

17. What is the difference, if any, between convection and heat radiation?

a. Thermal radiation is the transfer of heat from one place to another by the movement of fluids; convection is electromagnetic radiation emitted from all matter due to its possessing thermal energy.

b. Convection is the transfer of heat from one place to another by the movement of fluids; thermal radiation is nuclear energy emitted from all matter due to its possessing thermal energy.

c. Convection is the transfer of heat from one place to another by the movement of fluids; thermal radiation is electromagnetic radiation emitted from all matter due to its possessing thermal energy.

d. d. Convection is the transfer of heat from one place to another by the movement of fluids; thermal radiation is the barely detectable light emitted from all matter due to its possessing thermal energy.

18. Which of these statements about mechanical energy is/are true?

a. Mechanical energy is the energy that is possessed by an object due to its motion or due to its position.

b. Mechanical energy can be either kinetic energy (energy of motion) or potential energy (stored energy of position).

c. Objects have mechanical energy if they are in motion

d. All of the above.

19. _____ _____ is the energy released from an atom in nuclear reactions or by radioactive decay, especially the energy released in nuclear fission or nuclear fusion.

   a. Chemical energy
   b. Atomic energy
   c. Nuclear energy
   d. B and C are correct.

20. What is the difference between fusion and fission?

   a. Fusion splits a massive element into fragments, releasing energy in the process; fission joins two light elements, forming a more massive element and releasing energy in the process.

   b. Fission splits a massive element into fragments, retaining energy in the process; fusion joins two light elements, forming a more massive element and retaining energy in the process.

   c. Fission splits a massive element into fragments, releasing energy in the process; fusion joins two light elements, forming a more massive element and releasing energy in the process.

   d. There is very little difference between fusion and fission.

21. Which of the following statements about entropy is/are true?

   a. Ice melting provides an example in which entropy increases in small, thermodynamic system.

   b. Entropy change is the quantitative measure of the amount of energy that has flowed or how widely it has become spread out at a specific temperature.

   c. The concept of entropy is central to the second law of thermodynamics, which determines which physical processes can occur.

   d. All of the above.

22. Which of these statements about light energy is/are true?

a. Light consists of electromagnetic waves in the visible range.

b. The fundamental particle or quantum of light is a photon.

c. Both radiant and light energy are visible forms of energy.

d. A and B are true.

23. In the absence of magnetic or electric fields cathode rays _____.

a. Do not exist

b. Travel in straight lines

c. Become positively charged

d. Bend toward a light source

24. When pressure on a gas is reduced to half what happens to its volume?

a. The volume stays the same

b. The volume decreases

c. The volume rises then falls

d. The volume increases

Horseshoe Magnets

25. What shape are the magnetic field lines between the poles of a horseshoe magnet?

    a. Circular
    b. Oval
    c. Linear
    d. Rectangular

26. What happens if a horseshoe magnet is cut into two pieces as in the figure?

    a. Two unipolar magnets are created
    b. Two bipolar magnets are created
    c. The magnet is destroyed
    d. South pole becomes North and the North pole becomes South in the places where the N and S symbols are written

27. Which statement regarding horseshoe magnets is correct?

a. Horseshoe magnets are the strongest magnets available

b. Horseshoe magnets are used to show the geographic directions

c. North and South poles of horseshoe magnets lie in opposite directions

d. Horseshoe magnets are used to straighten the magnetic field lines

28. A compass is placed at four different positions near a horseshoe magnet as shown. In which position will the compass needle NOT rotate?

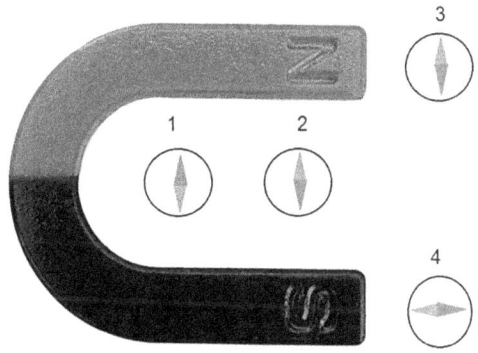

a. Position 1
b. Position 2
c. Position 3
d. Position 4

29. A horseshoe magnet is cut into four pieces as shown.

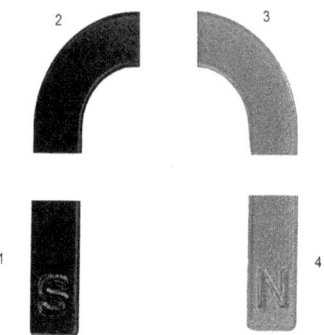

Which option shows the correct polarization of the third piece?

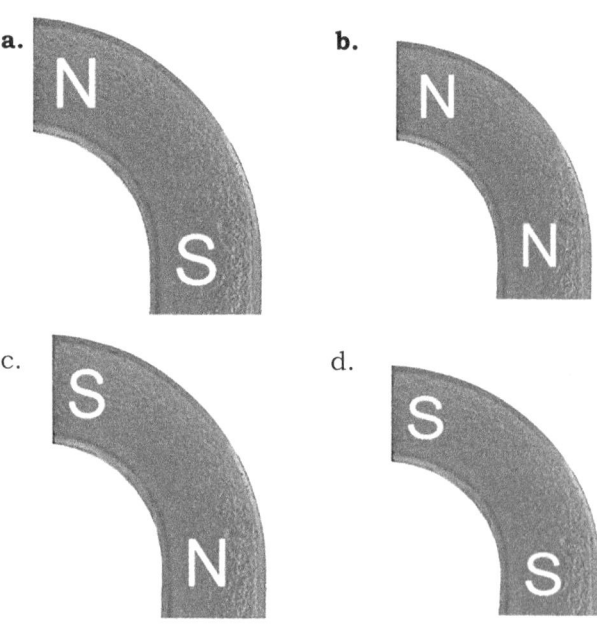

30. A thread is hanging a horseshoe magnet as shown. After making some rotations, the magnets stops moving at the position shown. Which number is the direction of the North Pole?

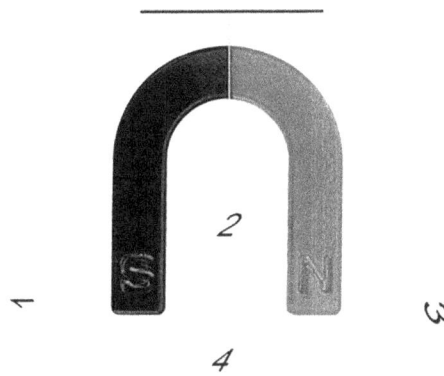

a. Direction 1
b. Direction 2
c. Direction 3
d. Direction 4

Answer Key

1. C
The work done against gravity for the increase in the gravitational potential energy of the objects shown. Since the gravitational potential energy of all objects is zero, this increase is equal to the actual gravitational potential energy when the objects are lifted to the shown position.

For the first object, we have

$W_1 = \Delta GPE_1 = GPE_{1\,final} = 3m * g * h = 3mgh$

For the second object, we have

$W_2 = \Delta GPE_2 = GPE_{2\,final} = m * g * 3h = 3mgh$

And for the third object, we have

$W_3 = \Delta GPE_3 = GPE_{3\,final} = 2m * g * 2h = 4mgh$

Thus, the work against gravity is the greatest in the third object.

2. D
Since the pushing force (and the motion of the object) is horizontal, there is no work done against gravity as it acts in the vertical (downward) direction.

(Do not confuse the work done against gravity with the work done to move the object. Here,, the result would be 20 · 10 = 200J.)

3. C
Choice A is incorrect. When we move an object downwards, we are acting in the direction of gravity.
Choice B is incorrect. As gravity acts vertically down, the horizontal motion is not affected by it. So, there is no work against gravity when moving an object horizontally.
Choice C is correct. When lifting an object upwards, we act against gravity (which acts downwards). As a result, the work done here is against gravity.

Choice D is incorrect. When an object is at rest, there is no work done on it.

4. D

The work done against gravity is the increase in the gravitational potential energy of the object. Since the initial gravitational potential energy the object is zero, this increase is equal to the actual gravitational potential energy when the object is lifted to the shown position.

The equation is

$$W_{against\ gravity} = \Delta GPE_{of\ object}$$

$$= GPE_{final}$$
$$= m * g * h$$
$$= 400 * 10 * 15$$
$$= 60,000 J$$

5. B

If there is no friction, the object's weight does not affect its horizontal motion. Thus, we need a minimum force (slightly greater than zero) to move it on a horizontal plane a distance d.

6. A

When considering the work done against gravity to lift an object through the stairs, the horizontal displacement is not counted. Thus, the only relevant distance is the vertical one (the increase in height).

In the figure, we see that the path (hypotenuse) is 10 m long and the horizontal displacement (leg) is 8m. Using the Pythagorean Theorem, we can find the increase in height:

$$h^2 = \sqrt{(10^2 - 8^2)} = \sqrt{(100 - 64)} = \sqrt{36} = 6m$$

The work done against gravity is the increase in the gravitational potential energy of the object. Since the initial gravitational potential energy the object is zero, this increase is equal to the actual gravitational potential energy when the object is lifted to the shown position.

The equation is,

$W_{against\ gravity} = \Delta GPE_{of\ object}$

$= GPE_{final}$
$= (m * g) * h$
$= F * h$
$= 100 * 6$
$= 600J$

7. B
Choice A is incorrect. When you throw an object downward, the work done is minimal as you are acting in the direction of gravity. Here, most of the work on the object is done by gravity.
Choice B is correct. The work done against gravity when throwing an object upwards is equal to the change in the gravitational potential energy of the object.
Choice C is incorrect. The work done against gravity when you push an object horizontally is zero.
Choice D is incorrect. In both cases, the work is zero as there is no motion.

8. B
Unless acted on by a force, a body in motion will change direction and gradually slow until it eventually stops.

This answer is related to Newton's 1st law of motion, which states that, unless acted on by a force, a body at rest stays at rest, and a moving body continues moving at the same speed in a straight line.

9. A
The formula for acceleration = $A = (V_f - V_0)/t$
so A = (10 m/sec - 0 m/sec)/20 sec = 0.5 m/sec^2

10. B
Speed = (total distance traveled)/(total time taken)
1000/5 = 200 meters per second

11. A
Force = Mass times Acceleration Measured in Newtons.
F = 2000 kg X 3 m/sec^2 = 6000 N

12. C
Electric potential is a fundamental interaction between the magnetic field and the presence and motion of an electric charge.

Electric potential is the capacity of an electric field to do work on an electric charge, typically measured in volts, while electromagnetism is a fundamental interaction between the magnetic field and the presence and motion of an electric charge

13. D
An ohm (Ω) is a unit of electrical voltage is not true.
Note: An ohm is a unit of electrical resistance.

14. D
The property of a conductor that restricts its internal flow of electrons is resistance.

15. C
In physics, friction is the force that opposes the relative motion of two bodies in contact.

16. B
Kinetic energy is the energy of a body that results from motion while potential energy is the energy possessed by an object by virtue of its position or state, e.g., as in a compressed spring.

17. C
Convection is the transfer of heat from one place to another by the movement of fluids; thermal radiation is electromagnetic radiation emitted from all matter due to its possessing thermal energy.

Note: In physics, the term "fluid" means any substance that deforms under shear stress; it includes liquids, gases, plasmas, and some plastic solids. Sunlight is solar electromagnetic radiation generated by the hot plasma of the Sun, and this thermal radiation heats the Earth.

18. A
All of the statements are true.
Note: Objects also have mechanical energy if they are at some position relative to a zero potential energy position,

for example, a brick held at a vertical position above the ground.

19. D
Atomic energy and nuclear energy are the energy released from an atom in nuclear reactions or by radioactive decay, especially the energy released in nuclear fission or nuclear fusion.

20. C
Fission splits a massive element into fragments, releasing energy in the process; fusion joins two light elements, forming a more massive element and releasing energy in the process.

21. D
All of the statements are true.

22. D
Note: Light energy is the only visible form of energy. A light bulb is a device that uses electrical energy to create electromagnetic energy in the form (in part) of visible light and heat.

23. B
Cathode rays (also called an electron beam or e-beam) are streams of electrons observed in vacuum tubes. They travel in straight lines through the empty tube. The voltage applied between the electrodes accelerates these low mass particles to high velocities.

24. D
Boyle's law (sometimes referred to as the Boyle-Mariotte law) is one of many gas laws and a special case of the ideal gas law. Boyle's law describes the inversely proportional relationship between the absolute pressure and volume of a gas, if the temperature is kept constant within a closed system.

Horseshoe Magnets

25. C
The field lines between the poles of a horseshoe magnet are linear. This is an advantage of horseshoe magnets (rectification of magnetic field lines).

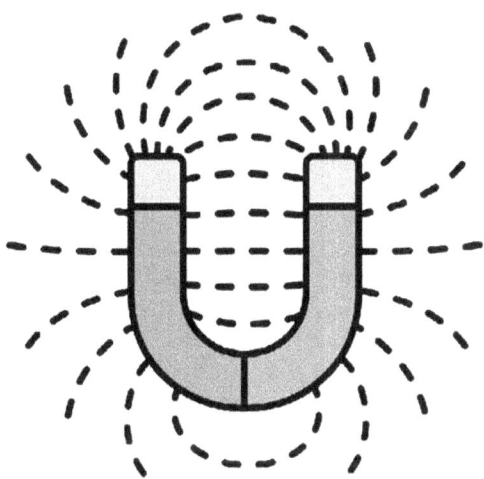

26. B
When a magnet is cut into two pieces, two new magnets with both poles are created. The new polarization occurs at the cutting point, while the other extremities preserve their previous polarization.

27. D
Choice A is incorrect. It is not the shape of magnets the factor which determined their strength.
Choice B is incorrect. Bar magnets, (not horseshoe) are used to show the geographic direction in compasses as their poles are in opposite direction.
Choice C is incorrect. North and South poles of horseshoe magnets lie in the same direction, not opposite.
Choice D is correct. Horseshoe magnets are used to straighten the magnetic field lines (in the space between the poles).

28. B
The universal rule of magnet states: "like poles repel each other while unlike poles attract each other. Thus, the only stable position in the figure is 2 as in that position unlike poles are facing (attracting) each other.

29. C
When a magnet is cut into two pieces, two new magnets with both poles are created. The new polarization occurs at the cutting point, while the other extremities preserve their previous polarization.

Here, the fourth piece remains N where it was before and becomes S at the cutting position. Therefore, the third piece becomes N in the lower side and N in the upper one.

30. C
The N-pole of a magnet is otherwise known as North seeking pole as it is directed due North pole of the Earth.

SPATIAL RELATIONS

Folding and Rotating

1. Ⓐ Ⓑ Ⓒ Ⓓ
2. Ⓐ Ⓑ Ⓒ Ⓓ
3. Ⓐ Ⓑ Ⓒ Ⓓ
4. Ⓐ Ⓑ Ⓒ Ⓓ
5. Ⓐ Ⓑ Ⓒ Ⓓ
6. Ⓐ Ⓑ Ⓒ Ⓓ
7. Ⓐ Ⓑ Ⓒ Ⓓ
8. Ⓐ Ⓑ Ⓒ Ⓓ
9. Ⓐ Ⓑ Ⓒ Ⓓ
10. Ⓐ Ⓑ Ⓒ Ⓓ
11. Ⓐ Ⓑ Ⓒ Ⓓ
12. Ⓐ Ⓑ Ⓒ Ⓓ
13. Ⓐ Ⓑ Ⓒ Ⓓ
14. Ⓐ Ⓑ Ⓒ Ⓓ
15. Ⓐ Ⓑ Ⓒ Ⓓ
16. Ⓐ Ⓑ Ⓒ Ⓓ
17. Ⓐ Ⓑ Ⓒ Ⓓ
18. Ⓐ Ⓑ Ⓒ Ⓓ
19. Ⓐ Ⓑ Ⓒ Ⓓ
20. Ⓐ Ⓑ Ⓒ Ⓓ
21. Ⓐ Ⓑ Ⓒ Ⓓ
22. Ⓐ Ⓑ Ⓒ Ⓓ
23. Ⓐ Ⓑ Ⓒ Ⓓ
24. Ⓐ Ⓑ Ⓒ Ⓓ
25. Ⓐ Ⓑ Ⓒ Ⓓ
26. Ⓐ Ⓑ Ⓒ Ⓓ
27. Ⓐ Ⓑ Ⓒ Ⓓ
28. Ⓐ Ⓑ Ⓒ Ⓓ
29. Ⓐ Ⓑ Ⓒ Ⓓ
30. Ⓐ Ⓑ Ⓒ Ⓓ
31. Ⓐ Ⓑ Ⓒ Ⓓ
32. Ⓐ Ⓑ Ⓒ Ⓓ
33. Ⓐ Ⓑ Ⓒ Ⓓ
34. Ⓐ Ⓑ Ⓒ Ⓓ
35. Ⓐ Ⓑ Ⓒ Ⓓ
36. Ⓐ Ⓑ Ⓒ Ⓓ
37. Ⓐ Ⓑ Ⓒ Ⓓ
38. Ⓐ Ⓑ Ⓒ Ⓓ
39. Ⓐ Ⓑ Ⓒ Ⓓ
40. Ⓐ Ⓑ Ⓒ Ⓓ

1. When the two longest sides touch what will the shape be?

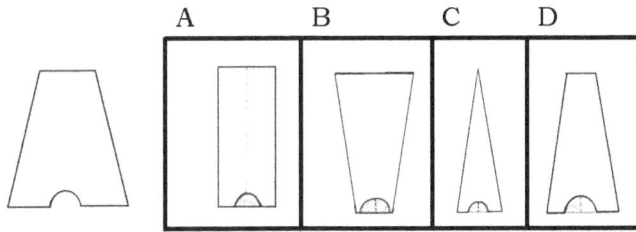

2. When folded, what pattern is possible?

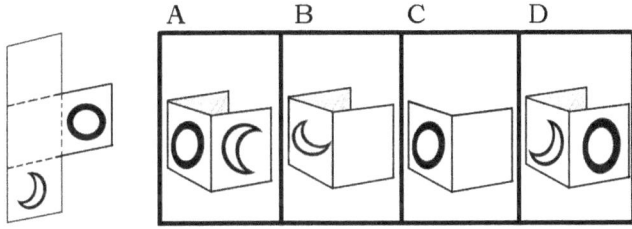

3. When folded into a loop, what will the strip of paper look like?

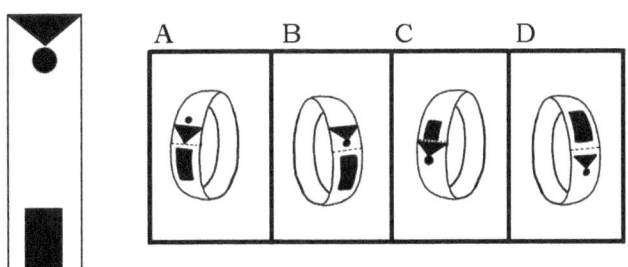

4. Which of the choices is the same pattern at a different angle?

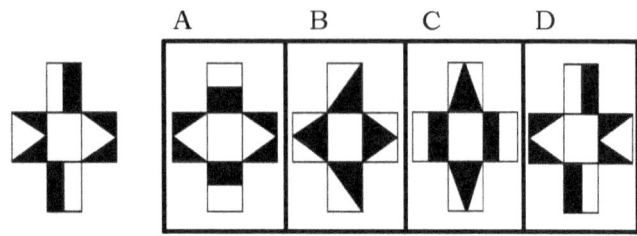

5. When put together, what 3-dimensional shape will you get?

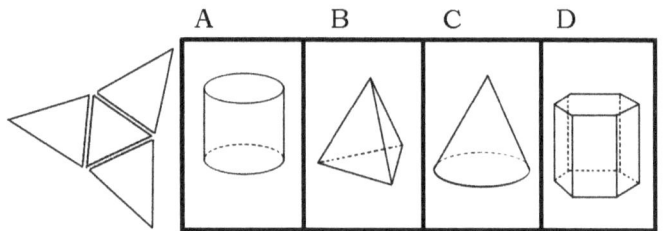

6. When folded, what pattern is possible?

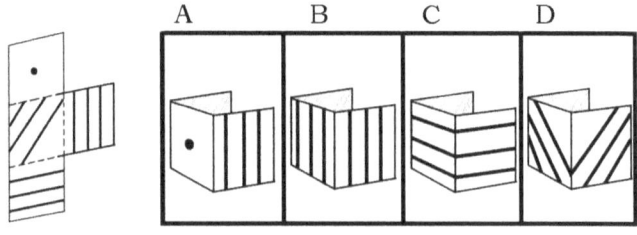

7. When folded into a loop, what will the strip of paper look like?

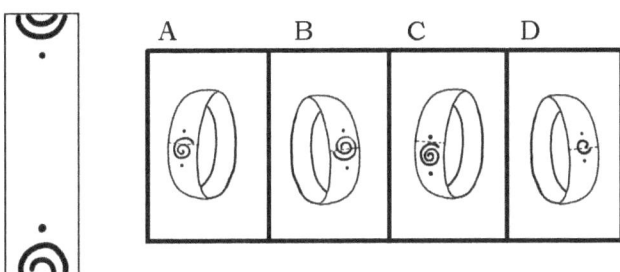

8. Which of the choices is the same pattern at a different angle?

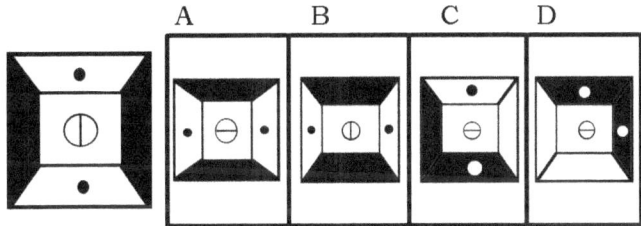

9. When folded, which shape will you get?

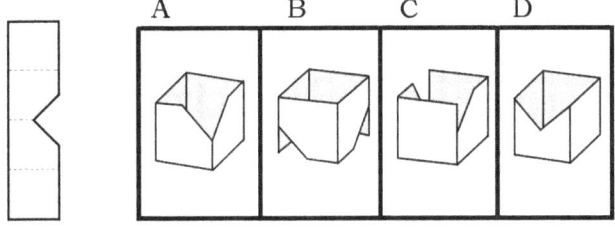

10. When folded, what pattern is possible?

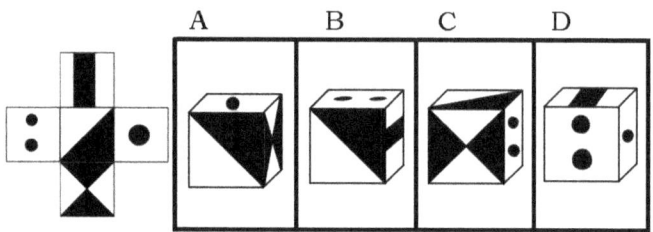

11. When folded, which shape is possible?

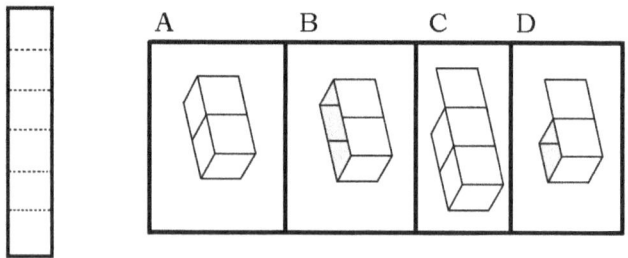

12. When folded, what pattern is possible?

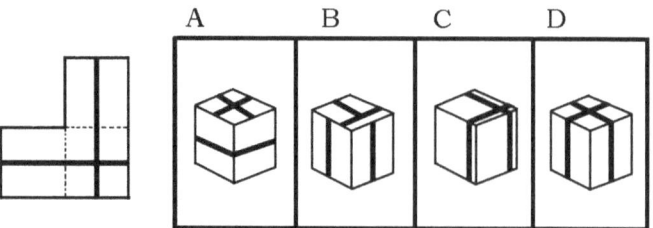

13. When folded into a loop, what will the strip of paper look like?

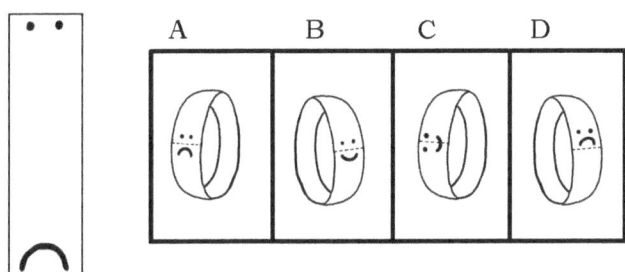

14. Which of the choices is the same pattern at a different angle?

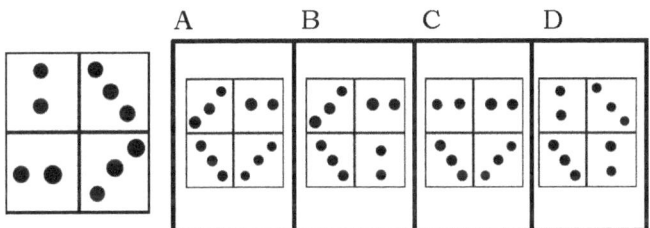

15. When folded along the dotted lines, which shape will you get?

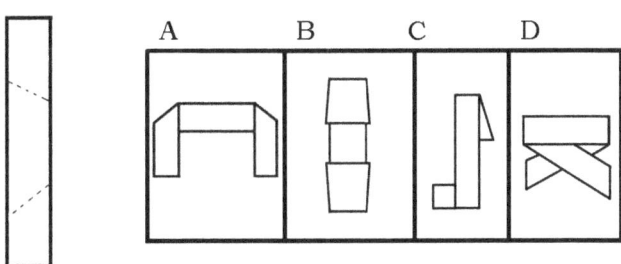

16. When folded, what pattern is possible?

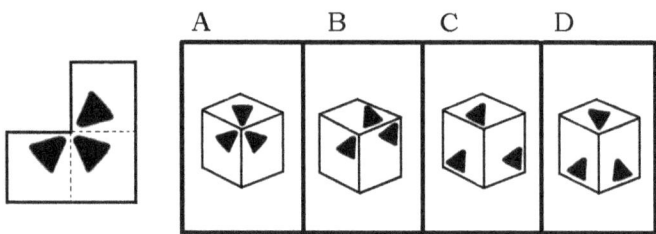

17. When folded into a loop, what will the strip of paper look like?

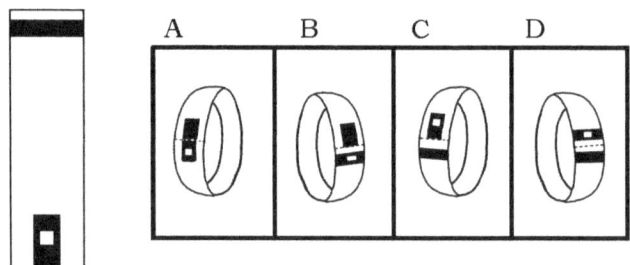

18. Which of the choices is the same pattern at a different angle?

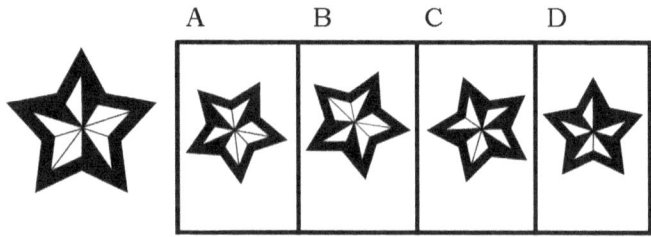

19. When folded along the dotted line, which shape will you get?

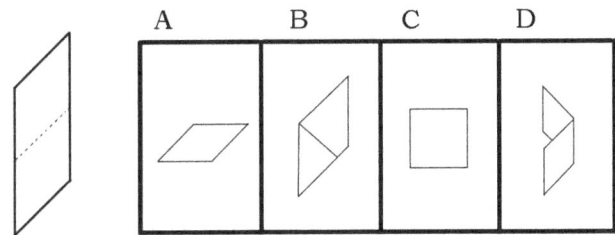

20. When folded, what pattern is possible?

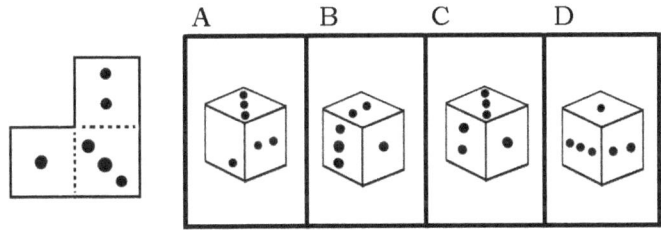

21. When folded, what pattern is possible?

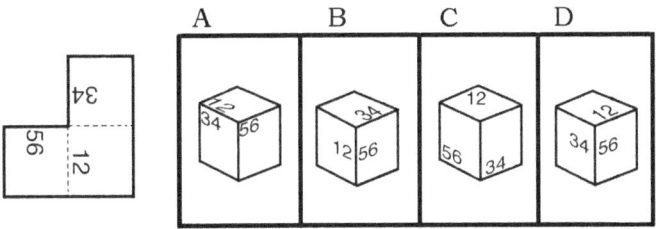

22. When folded into a loop, what will the strip of paper look like?

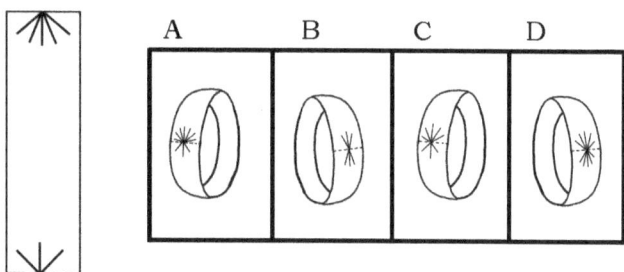

23. Which of the choices is the same pattern at a different angle?

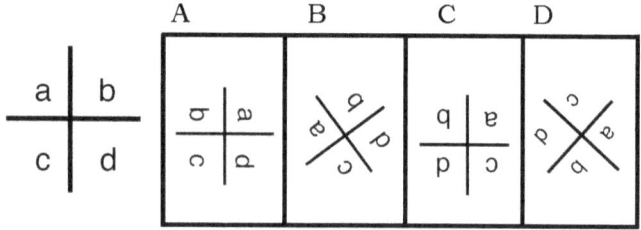

24. When folded, what pattern is possible?

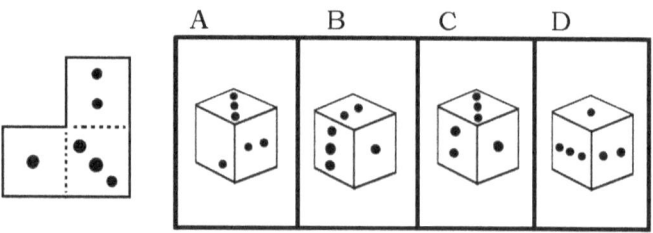

25. Which of the choices is the same pattern at a different angle?

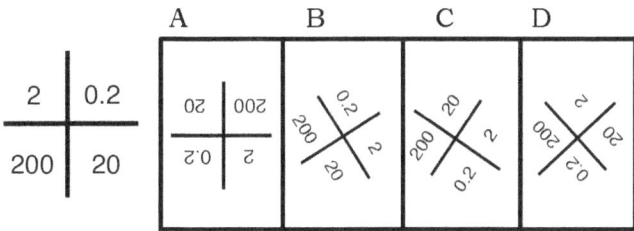

26. When folded along the dotted lines, which shape will you get?

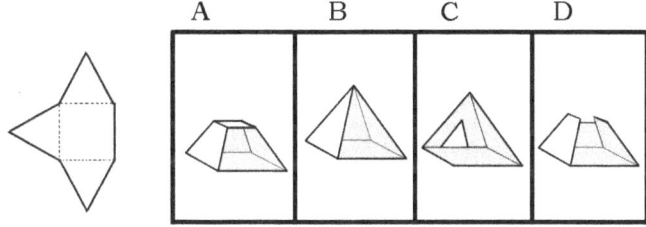

27. When folded, what pattern is possible?

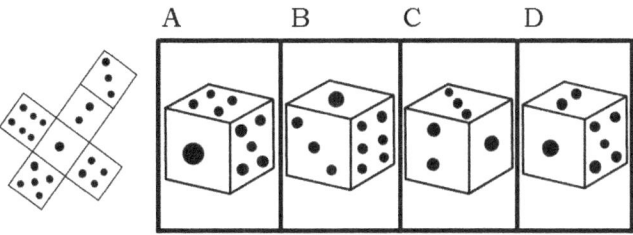

28. When folded into a loop, what will the strip of paper look like?

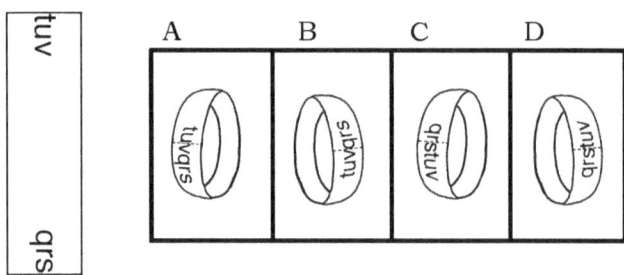

29. Which of the choices is the same pattern at a different angle?

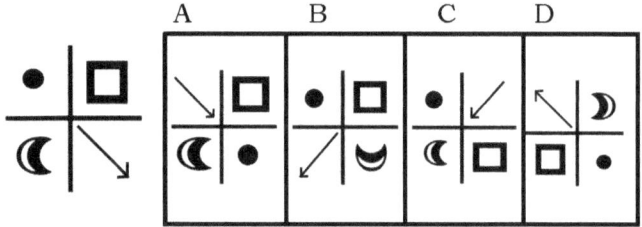

30. When put together, what 3-dimensional shape will you get?

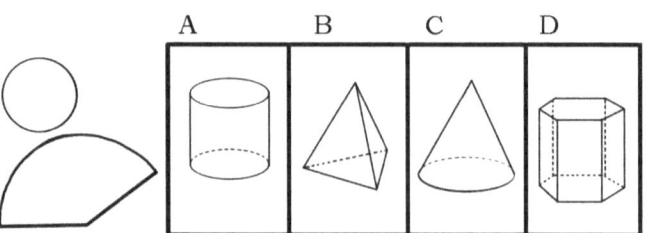

31. When folded, what pattern is possible?

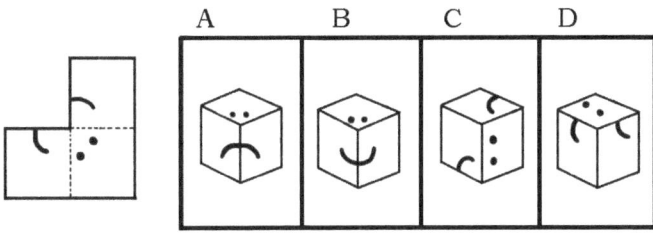

32. When folded, what pattern is possible?

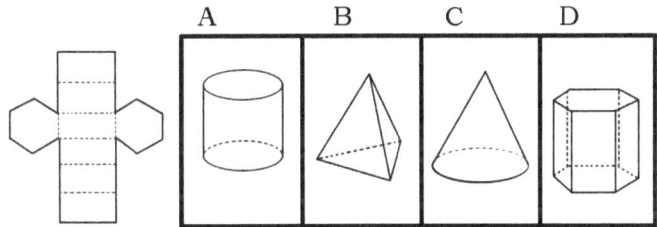

33. Which of the choices is the same pattern at a different angle?

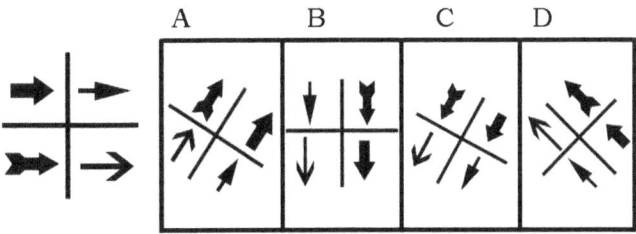

34. When put together, what 3-dimensional shape will you get?

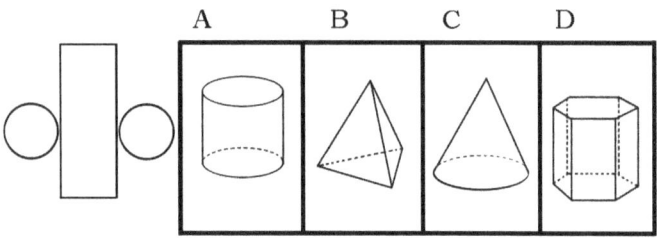

35. When folded into a loop, what will the strip of paper look like?

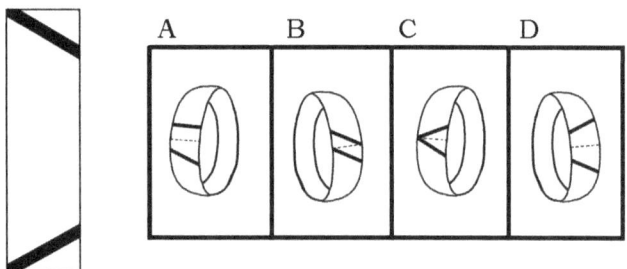

36. Which of the choices is the same pattern at a different angle?

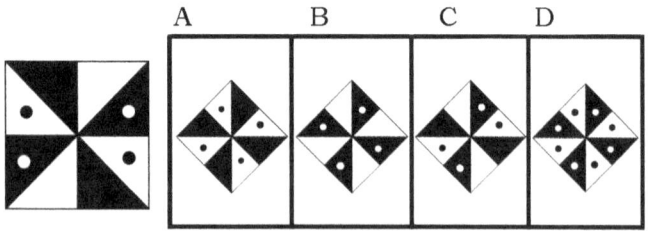

37. When put together, what 3-dimensional shape will you get?

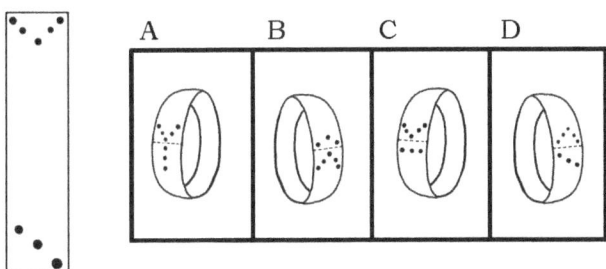

38. When folded into a loop, what will the strip of paper look like?

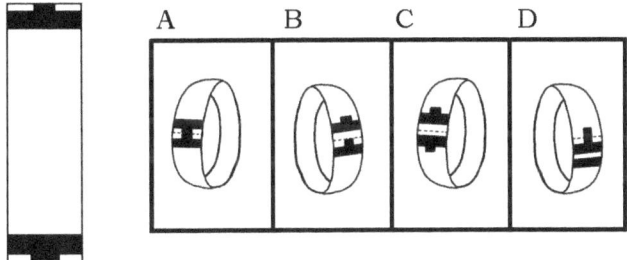

39. Which of the choices is the same pattern at a different angle?

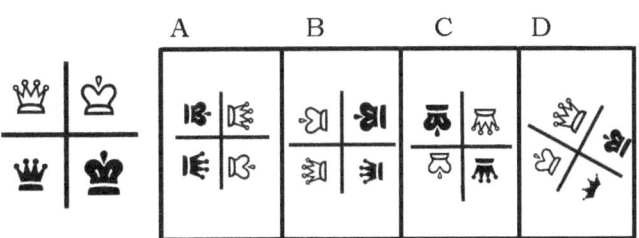

40. When folded into a loop, what will the strip of paper look like?

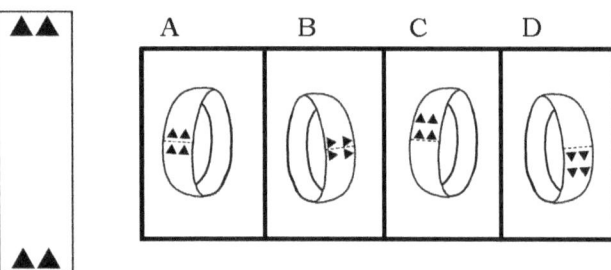

Answer Key

Folding and Rotating

1. D
2. A
3. C
4. B
5. B
6. C
7. B
8. A
9. A
10. A
11. B
12. D
13. B
14. B
15. A
16. A
17. C
18. B
19. D
20. C
21. D
22. C
23. D
24. C
25. A
26. B
27. A
28. D
29. D

30. C
31. B
32. D
33. C
34. A
35. C
36. C
37. D
38. A
39. B
40. A

Assembly

	A	B	C	D
1	○	○	○	○
2	○	○	○	○
3	○	○	○	○
4	○	○	○	○
5	○	○	○	○
6	○	○	○	○
7	○	○	○	○
8	○	○	○	○
9	○	○	○	○
10	○	○	○	○
11	○	○	○	○
12	○	○	○	○
13	○	○	○	○
14	○	○	○	○
15	○	○	○	○

1. Which figure represents the assembly of the following pieces?

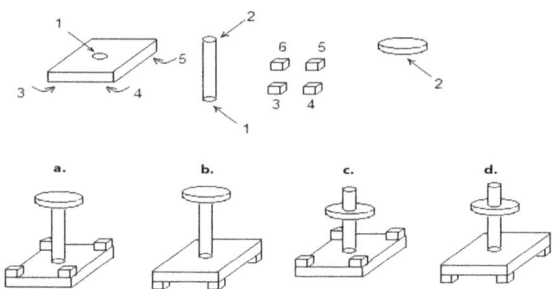

2. Which figure represents the assembly of the following pieces?

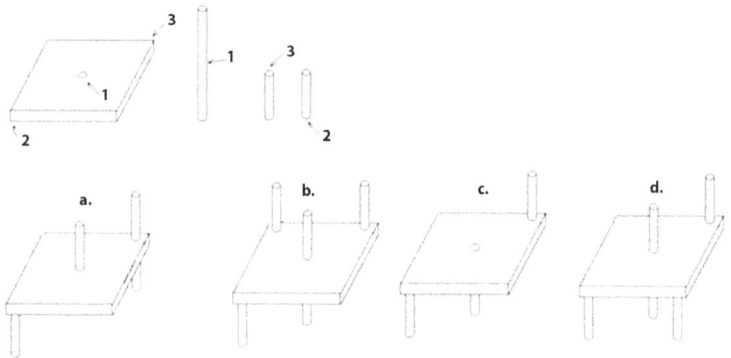

3. Which figure represents the assembly of the following pieces?

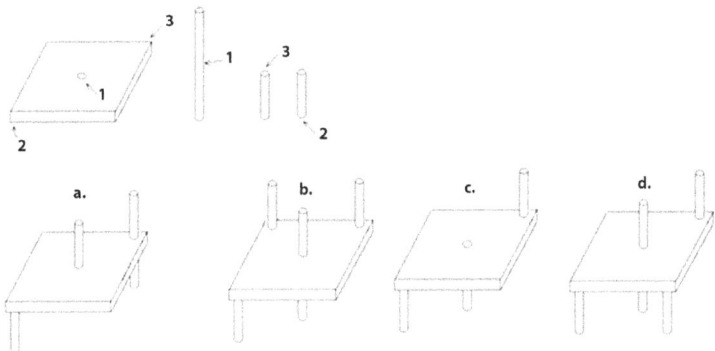

4. Which figure represents the assembly of the following pieces?

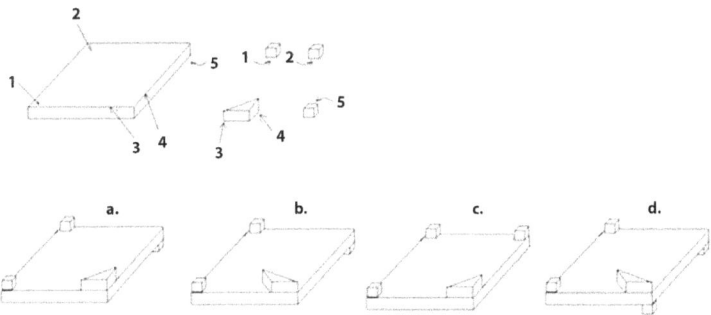

5. Which figure represents the assembly of the following pieces?

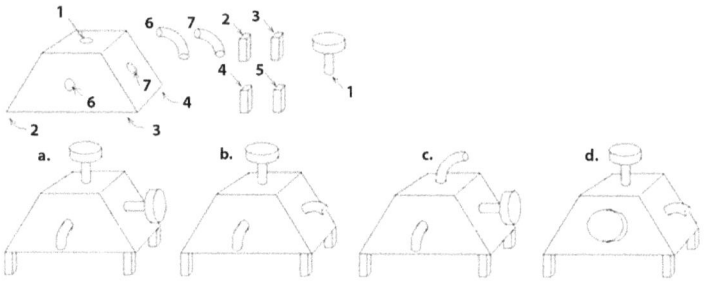

6. Which figure represents the assembly of the following pieces?

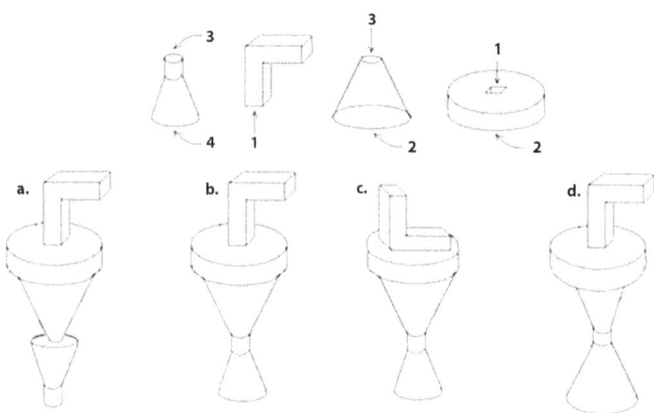

7. Which figure represents the assembly of the following pieces?

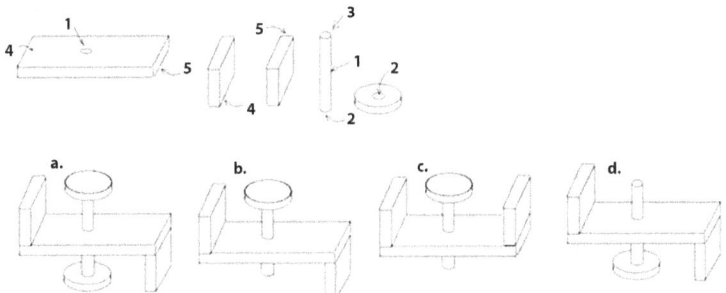

8. Which figure represents the assembly of the following pieces?

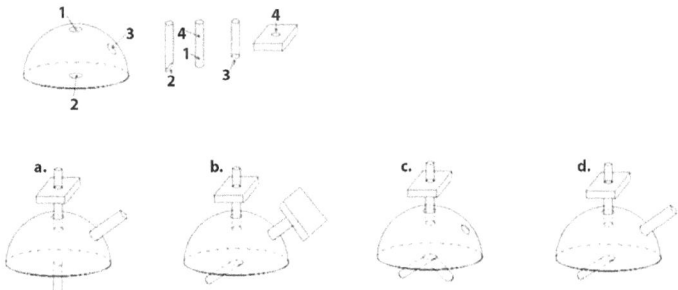

9. Which figure represents the assembly of the following pieces?

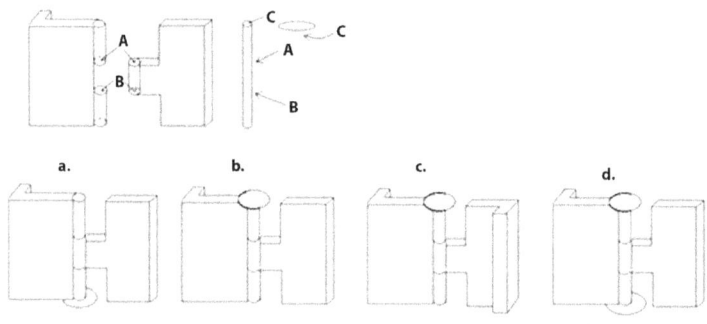

10. Which figure represents the assembly of the following pieces?

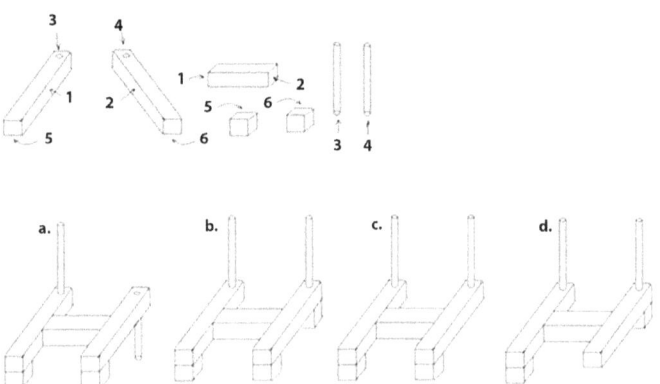

11. Which figure represents the assembly of the following pieces?

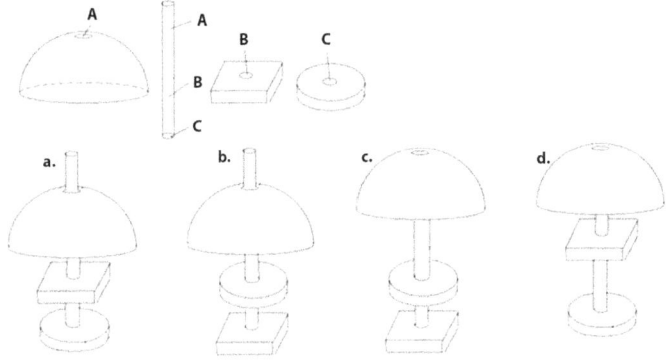

12. Which figure represents the assembly of the following pieces?

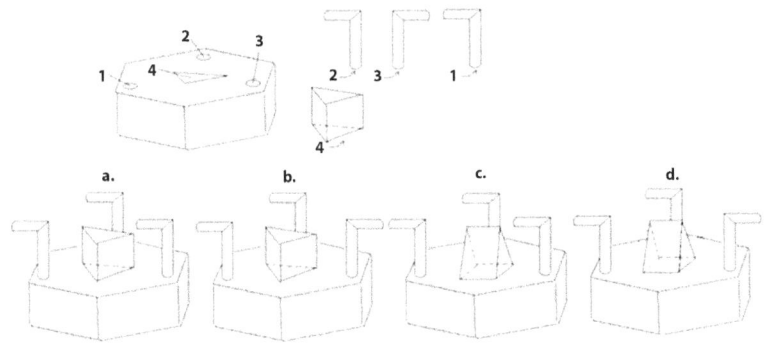

13. Which figure represents the assembly of the following pieces?

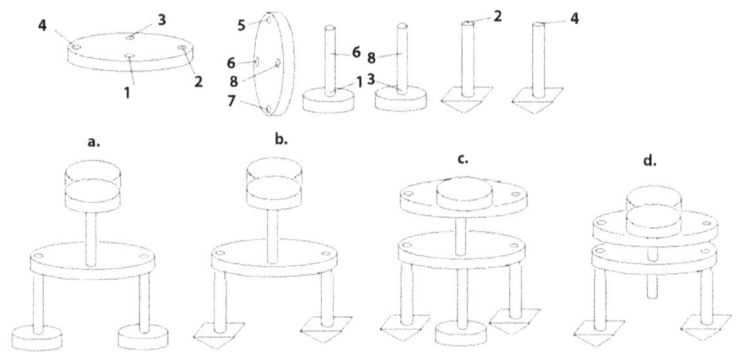

14. Which figure represents the assembly of the following pieces?

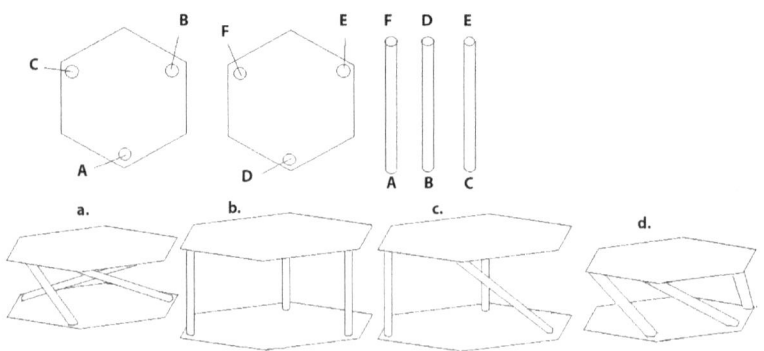

15. Which figure represents the assembly of the following pieces?

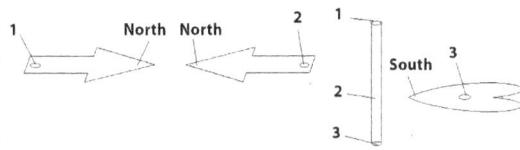

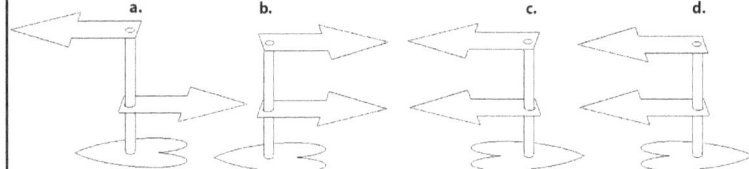

Answer Key

1. B
If two pieces have the same number at the position shown, it means that point is a junction point. Here, the cylindrical rod is at center of the rectangular platform, the small cubes are below the platform at its edges and the disc is above the rod.

2. D
If two pieces have the same number at the position shown, it means that point is a junction point. Here, the long rod is half above and half below the rectangular platform, and the short rods are one above and the other below the platform.

3. D
If two pieces have the same number at the position shown, it means that point is a junction point. Here, the rod connects the two wide wavy pieces and the two T-shapes are at the edges of the central rod.

4. A
If two pieces have the same number at the position shown, it means that point is a junction point. Here, all the small shapes are on the rectangular platform, where the triangular shape is on left-bottom corner and the three small cubes are at the other corners of the platform.

5. B
If two pieces have the same number at the position shown, it means that point is a junction point. Here, the hoses are at the central holes of the lateral faces of the platform, the screw-like shape is on top of the platform and the small cuboids act as legs.

6. B
If two pieces have the same number at the position shown, it means that point is a junction point. Following this rule, here will find that the correct assembly is shown at A.

7. D
If two pieces have the same number at the position shown, it means that point is a junction point. Here, the long rod is half above and half below the rectangular platform; the disc is at bottom of the rod, and the small rectangular shapes are in a vertical position at the extremities of the big rectangular platform, where the first is below the platform (the one on the right) and the other is above it (the one on the left).

8. D
If two pieces have the same number at the position shown, it means that point is a junction point. Here, the hose with one diagonal end is at the bottom of the half-sphere, while the hammer-like shape is on top.

9. B
If two pieces have the same number at the position shown, it means that point is a junction point. Here, the screw-like shape connects the two rectangular shapes with the head up.
Feedback for choice C
You may consider choice C, but one of the rectangular shapes is different from those given in the question.

10. C
If two pieces have the same number at the position shown, it means that point is a junction point. Here, there is a H-shaped object placed in the horizontal position and two vertical rods on the upper end of the H-shaped object.

11. A
If two pieces have the same number at the position shown, it means that point is a junction point. Here, there is a central rod, a half-sphere near the top of the rod, a rectangular-shaped object below the half of the rod and a disc at bottom of the rod.

12. B
If two pieces have the same number at the position shown, it means that point is a junction point. Following this rule, and maintaining the direction of the given shapes, you will notice that only B matches the description.

13. D

If two pieces have the same number at the position shown, it means that point is a junction point. Here, the screw-like shapes are placed vertically up, while the hammer-like shapes are vertically down. Also, there are two large discs at center of the platform.

14. A

If two pieces have the same number at the position shown, it means that point is a junction point. Here, the legs connect the positions shown in the upper platform with some points, which are shifted by two positions anti-clockwise.

15. D

If two pieces have the same number at the position shown, it means that point is a junction point. Here, the arrows and the upper part of the heart-shape must be in the same direction.

Line Following

	A	B	C	D
1	○	○	○	○
2	○	○	○	○
3	○	○	○	○
4	○	○	○	○
5	○	○	○	○
6	○	○	○	○
7	○	○	○	○
8	○	○	○	○
9	○	○	○	○
10	○	○	○	○
11	○	○	○	○
12	○	○	○	○
13	○	○	○	○
14	○	○	○	○
15	○	○	○	○
16	○	○	○	○
17	○	○	○	○
18	○	○	○	○
19	○	○	○	○
20	○	○	○	○

Questions 1 - 5 refer to the following diagram

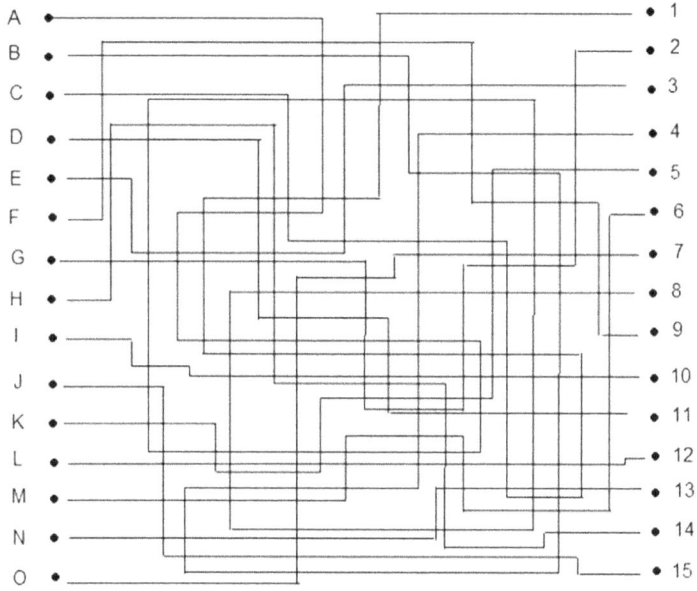

1. Which of the following is correct for the matching lines shown in the figure?

a. A-1   E-5   I-14   L-12
b. A-8   E-3   I-10   L-13
c. A-8   E-3   I-10   L-12
d. A-5   E-11   I-2   L-7

2. Which of the following is correct for the matching lines shown in the figure?

a. B-4   F-9   J-15   M-6
b. B-4   F-3   J-15   M-13
c. B-6   F-9   J-12   M-7
d. B-8   F-11   J-2   M-6

3. **Which of the following is correct for the matching lines shown in the figure?**

   a. C-4   G-9   K-15   N-6
   b. C-7   G-10   K-5   N-3
   c. C-1   G-2   K-12   N-7
   d. C-1   G-2   K-5   N-13

4. **Which of the following is correct for the matching lines shown in the figure?**

   a. D-5   H-10   L-11   O-13
   b. D-11   H-14   L-12   O-7
   c. D-5   H-7   L-10   O-11
   d. D-3   H-6   L-8   O-12

5. **Which of the following is correct for the matching lines shown in the figure?**

   a. A-4   C-11   M-12   L-13
   b. A-1   C-8   M-14   L-5
   c. A-8   C-1   M-6   L-12
   d. A-2   C-7   M-8   L-10

Questions 6 - 10 refer to the following diagram

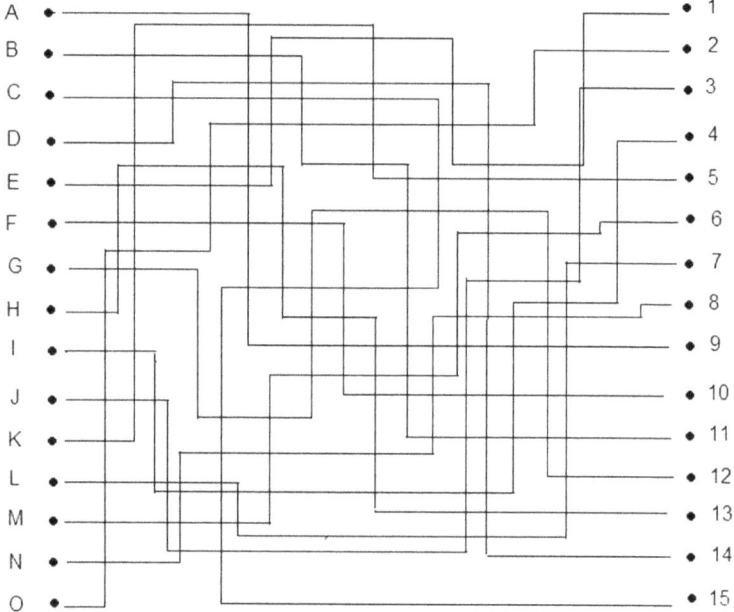

6. Which of the following is correct for the matching lines shown in the figure?

    a. D-4   G-12   I-14   K-5

    b. D-14   G-12   I-4   K-5

    c. D-4   G-10   I-6   K-12

    d. D-2   G-7   I-9   K-10

7. Which of the following is correct for the matching lines shown in the figure?

    a. A-9   E-1   M-6   N-8

    b. A-7   E-5   M-12   N-6

    c. A-9   E-1   M-6   N-12

    d. A-8   E-7   M-10   N-1

8. Which of the following is correct for the matching lines shown in the figure?

   a. B-4   C-8   J-1   F-7
   b. B-5   C-7   J-12   F-6
   c. B-11   C-1   J-6   F-13
   d. B-11   C-15   J-3   F-10

9. Which of the following is correct for the matching lines shown in the figure?

   a. D-4   H-13   L-2   O-7
   b. D-14   H-13   L-7   O-2
   c. D-14   H-2   L-7   O-13
   d. D-13   H-14   L-2   O-7

10. Which of the following is correct for the matching lines shown in the figure?

   a. B-11   C-15   G-12   H-14
   b. B-1   C-5   G-2   H-4
   c. B-14   C-12   G-11   H-15
   d. B-13   C-14   G-2   H-6

Questions 11 - 15 refer to the following diagram

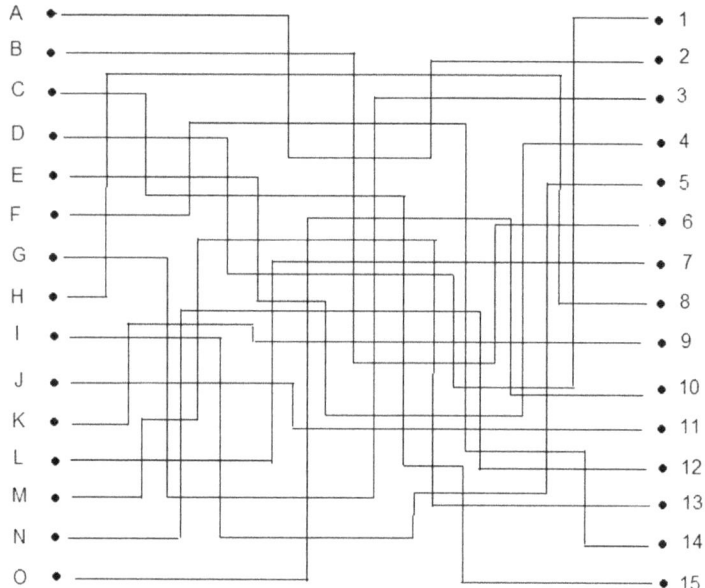

11. Which of the following is correct for the matching lines shown in the figure?

a.	A-2	B-5	C-6	D-1
b.	A-2	B-6	C-15	D-1
c.	A-4	B-13	C-2	D-5
d.	A-7	B-11	C-8	D-6

12. Which of the following is correct for the matching lines shown in the figure?

a.	A)	E-4	F-14	G-3	H-8
b.	B)	E-3	F-6	G-12	H-5
c.	C)	E-8	F-13	G-7	H-9
d.	D)	E-7	F-11	G-6	H-8

LINE FOLLOWING

13. Which of the following is correct for the matching lines shown in the figure?

a. I-5	J-7	K-9	L-11
b. I-5	J-7	K-11	L-9
c. I-5	J-11	K-9	L-7
d. I-7	J-11	K-8	L-9

14. Which of the following is correct for the matching lines shown in the figure?

a. M-13	N-12	O-10	E-4
b. M-12	N-13	O-14	E-9
c. M-5	N-11	O-9	E-7
d. M-7	N-11	O-7	E-9

15. Which of the following is correct for the matching lines shown in the figure?

a. G-3	B-12	D-6	F-14
b. G-2	B-13	D-4	F-9
c. G-5	B-14	D-6	F-3
d. G-3	B-6	D-1	F-14

Questions 16 - 20 refer to the following diagram

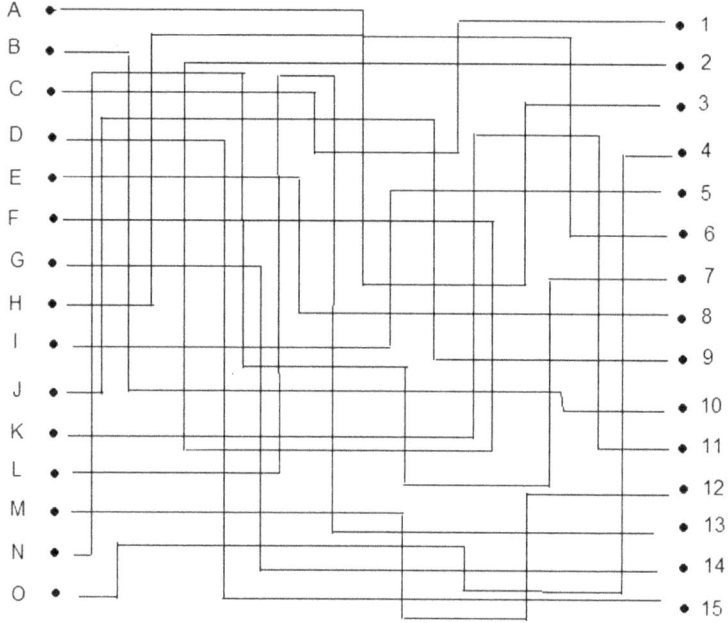

16. Which of the following is correct for the matching lines shown in the figure?

    a. A-3   B-10   C-1   D-15
    b. A-3   B-11   C-10   D-15
    c. A-3   B-13   C-12   D-5
    d. A-5   B-11   C- 3   D-15

17. Which of the following is correct for the matching lines shown in the figure?

    a. E-12   F-14   G-3   H-8
    b. E-13   F-6   G-12   H-6
    c. E-8   F-13   G-2   H-6
    d. E-8   F-2   G- 14   H-6

18. **Which of the following is correct for the matching lines shown in the figure?**

   a. I-5   J-7   K-9   L-11
   b. I-5   J-9   K-11   L-13
   c. I-5   J-11   K-9   L-13
   d. I-7   J-11   K-12   L-9

19. **Which of the following is correct for the matching lines shown in the figure?**

   a. M-13   N-12   O-4   L-7
   b. M-12   N-7   O-4   L-13
   c. M-15   N-11   O-13   L-4
   d. M-7   N-12   O-11   L-5

20. **Which of the following is correct for the matching lines shown in the figure?**

   a. C-1   F-2   D-15   M-12
   b. C-2   F-1   D-15   M-12
   c. C-5   F-12   D-2   M-3
   d. C-4   F-6   D-11   M-14

ANSWER KEY

Diagram for Questions 1 - 5

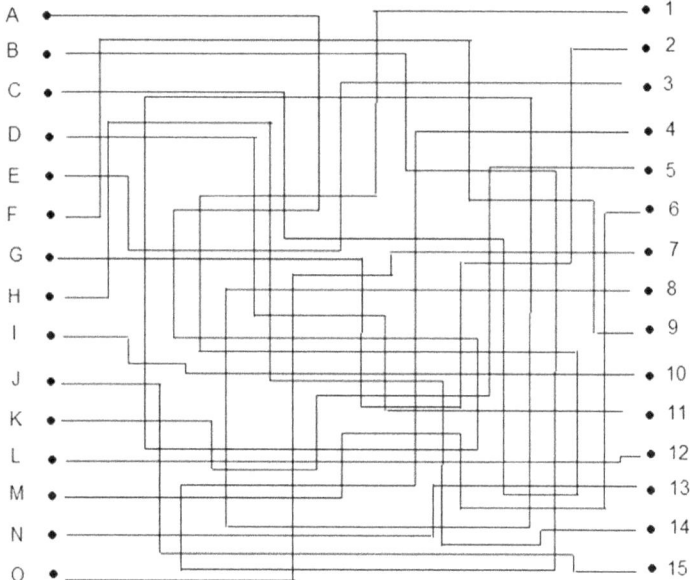

1. C
Carefully observing the lines, you will notice that the line starting from A ends at 8, the line starting from E ends at 3, the line starting from I ends at 10 and the line starting from L ends at 12.

2. A
Carefully observing the lines, you will notice that the line starting from B ends at 4, the line starting from F ends at 9, the line starting from J ends at 15 and the line starting from M ends at 6.

3. D
Carefully observing the lines, you will notice that the line starting from C ends at 1, the line starting from G ends at 2, the line starting from K ends at 5 and the line starting from N ends at 13.

4. B
Carefully observing the lines, you will notice that the line starting from D ends at 11, the line starting from H ends at 14, the line starting from L ends at 12 and the line starting from O ends at 7.

5. C
Carefully observing the lines, you will notice that the line starting from A ends at 11, the line starting from C ends at 1, the line starting from M ends at 6 and the line starting from L ends at 12.

Diagram for Questions 6 - 10

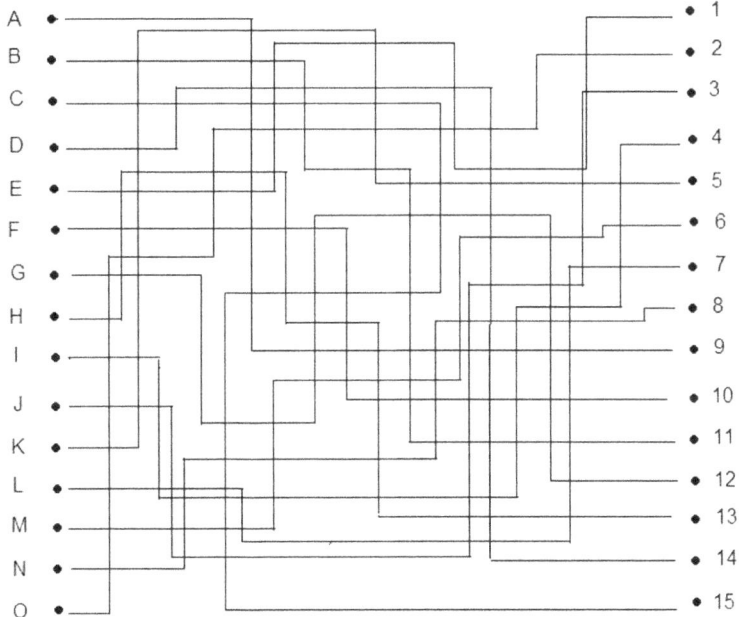

6. B
Carefully observing the lines, you will notice that the line starting from D ends at 14, the line starting from G ends at 12, the line starting from I ends at 4 and the line starting from K ends at 5.

7. A
Carefully observing the lines, you will notice that the line starting from A ends at 9, the line starting from E ends at 1, the line starting from M ends at 6 and the line starting from N ends at 8.

8. D
Carefully observing the lines, you will notice that the line starting from B ends at 11, the line starting from C ends at 15, the line starting from J ends at 3 and the line starting from F ends at 10.

9. B
Carefully observing the lines, you will notice that the line starting from D ends at 14, the line starting from H ends at 13, the line starting from L ends at 7 and the line starting from O ends at 2.

10. A
Carefully observing the lines, you will notice that the line starting from B ends at 11, the line starting from C ends at 15, the line starting from G ends at 12 and the line starting from H ends at 14.

Diagram for Questions 11 - 15

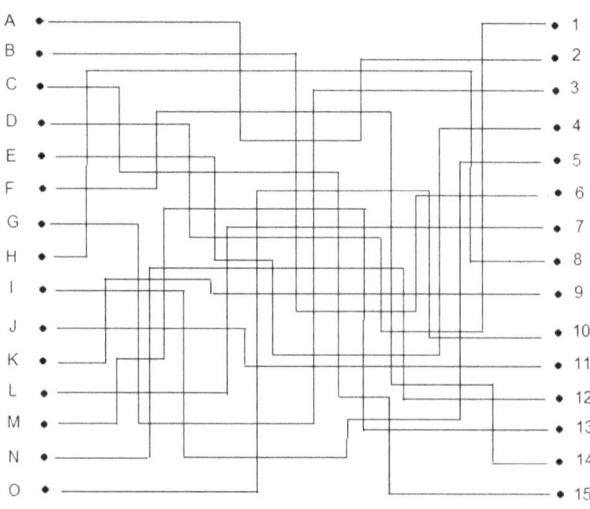

11. B
Carefully observing the lines, you will notice that the line starting from A ends at 2, the line starting from B ends at 6, the line starting from C ends at 15 and the line starting from D ends at 1.

12. A
Carefully observing the lines, you will notice that the line starting from E ends at 4, the line starting from F ends at 14, the line starting from G ends at 3 and the line starting from H ends at 8.

13. C
Carefully observing the lines, you will notice that the line starting from I ends at 5, the line starting from J ends at 11, the line starting from K ends at 9 and the line starting from L ends at 7.

14. A
Carefully observing the lines, you will notice that the line starting from M ends at 13, the line starting from N ends at 12, the line starting from O ends at 10 and the line starting from E ends at 4.

15. D
Carefully observing the lines, you will notice that the line starting from G ends at 3, the line starting from B ends at 6, the line starting from D ends at 1 and the line starting from F ends at 14.

Diagram for Questions 16 - 20

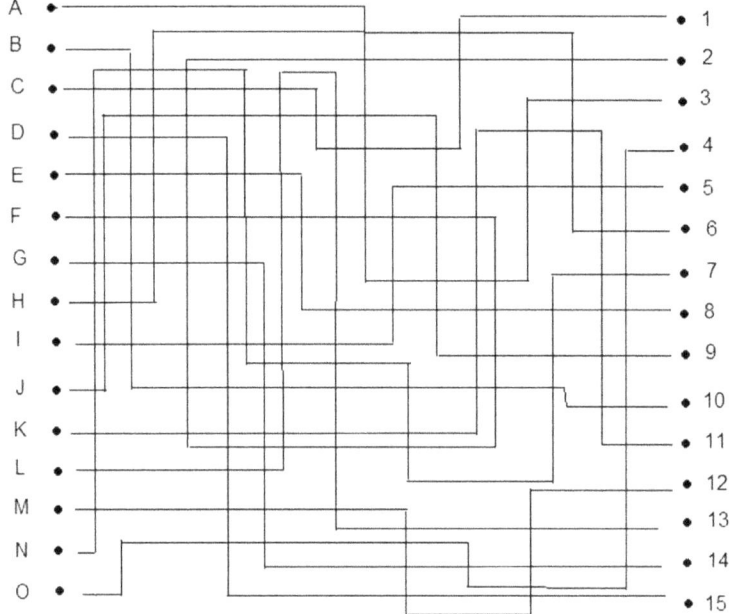

16. A
Carefully observing the lines, you will notice that the line starting from A ends at 3, the line starting from B ends at 10, the line starting from C ends at 1 and the line starting from D ends at 15.

17. D
Carefully observing the lines, you will notice that the line starting from E ends at 4, the line starting from F ends at 14, the line starting from G ends at 3 and the line starting from H ends at 8.

18. B
Carefully observing the lines, you will notice that the line starting from I ends at 5, the line starting from J ends at 9, the line starting from K ends at 11 and the line starting from L ends at 13.

19. B

Carefully observing the lines, you will notice that the line starting from M ends at 12, the line starting from N ends at 7, the line starting from O ends at 4 and the line starting from L ends at 13.

20. A

Carefully observing the lines, you will notice that the line starting from C ends at 1, the line starting from F ends at 2, the line starting from D ends at 15 and the line starting from M ends at 12.

Touching Blocks

	A	B	C	D
1	○	○	○	○
2	○	○	○	○
3	○	○	○	○
4	○	○	○	○
5	○	○	○	○
6	○	○	○	○
7	○	○	○	○
8	○	○	○	○
9	○	○	○	○
10	○	○	○	○
11	○	○	○	○
12	○	○	○	○
13	○	○	○	○
14	○	○	○	○
15	○	○	○	○

Questions 1 - 3 refer to the following diagram

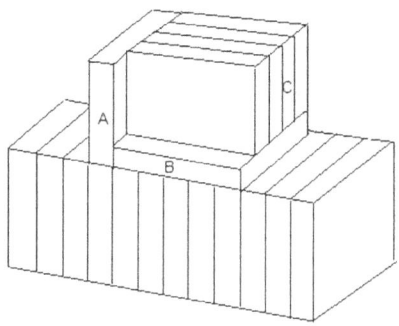

1. How many blocks is block A touching?

   a. 4
   b. 5
   c. 6
   d. 7

2. How many blocks is block B touching?

   a. 4
   b. 5
   c. 9
   d. 10

3. How many blocks is block C touching?

   a. 4
   b. 3
   c. 2
   d. 1

Questions 4 - 6 refer to the following diagram

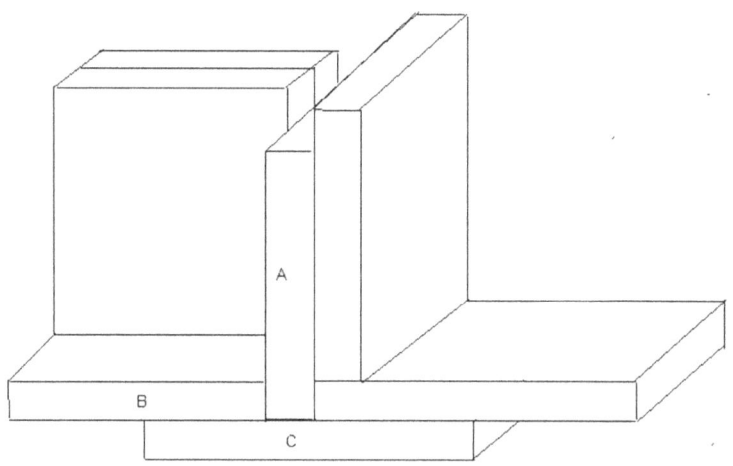

4. How many blocks is block A touching?

   a. 7
   b. 6
   c. 5
   d. 4

5. How many blocks is block B touching?

   a. 4
   b. 5
   c. 6
   d. 3

6. How many blocks is block C touching?

   a. 4
   b. 3
   c. 2
   d. 1

Questions 7 - 9 refer to the following diagram

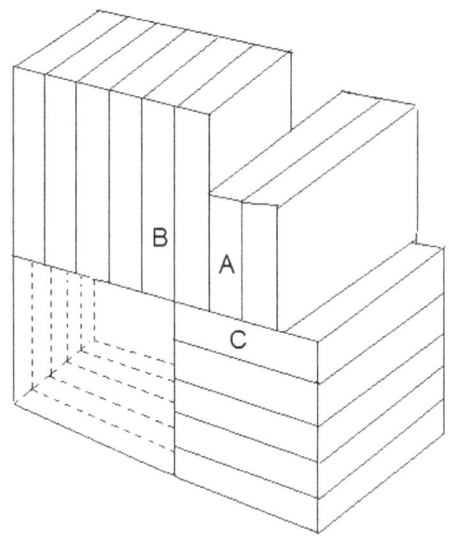

7. **How many blocks is block A in the figure touching? The blocks on the bottom-left side are transparent.**

    a. 6
    b. 5
    c. 4
    d. 3

8. **How many blocks is block B in the figure touching? The blocks on the bottom-left side are transparent.**

    a. 7
    b. 6
    c. 5
    d. 4

9. **How many blocks is block B in the figure touching? The blocks on the bottom-left side are transparent.**

    a. 10
    b. 9
    c. 8
    d. 7

Questions 10 - 12 refer to the following diagram

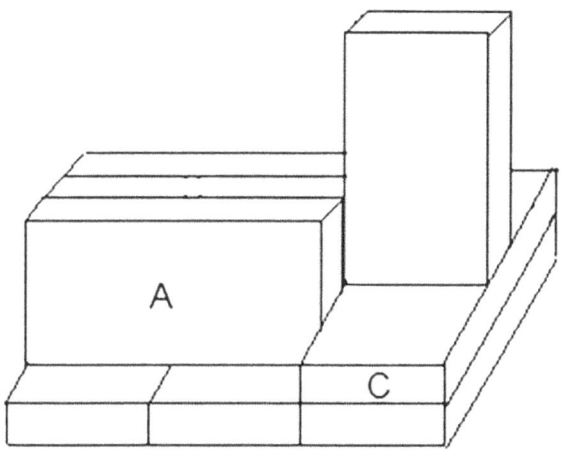

10. How many blocks is block A touching?

    a. 6
    b. 5
    c. 4
    d. 3

11. How many blocks is block B touching?

    a. 6
    b. 5
    c. 4
    d. 3

12. How many blocks is block C in the figure touching?

   a. 8
   b. 7
   c. 6
   d. 5

Questions 13 - 15 refer to the following diagram

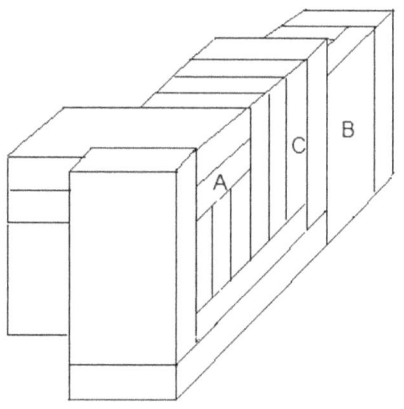

13. How many blocks is block A in the figure touching?

   a. 6
   b. 5
   c. 4
   d. 3

14. How many blocks is block B in the figure touching?

   a. 6
   b. 5
   c. 4
   d. 3

15. How many blocks is block C in the figure touching?

   a. 6
   b. 5
   c. 4
   d. 3

Answer Key

1. C
Block A touches 6 blocks (1 is below, 4 are lateral in vertical position and 1 is lateral in the horizontal position.

2. D
Block B touches 5 blocks below, 4 blocks above and one block laterally, i.e. in total 10 blocks.

3. A
Block C touches 4 blocks in total: 1 below, 2 blocks laterally on the wider side and one block laterally, in the narrower side.

4. B
Block A touches 6 blocks (1 is below, 2 are lateral in vertical position (narrow face), 2 are lateral in the horizontal position and one is lateral in the vertical position (wider side).

5. A
Block B touches 1 block below, 2 blocks above and one block laterally, i.e. in total 4 blocks.

6. B
Block C touches 3 blocks in total: all of them above it.

7. D
Block A touches 3 blocks in total: 2 are laterally placed and the other is below it.

8. A
Block B touches 5 blocks below and 2 laterally placed, i.e. in total 7 blocks.

9. B
Block C touches 1 blocks below, 5 laterally placed blocks, and 3 other blocks above it, i.e. 9 blocks in total.

10. C
Block A touches 4 blocks in total: 1 laterally placed, 2 below and 1 behind it.

11. B
Block B touches 5 blocks in total: 1 laterally placed, 2 below, 1 before and 1 behind it.

12. D
Block C touches 5 blocks in total: 3 laterally placed, 1 below and one above it.

13. A
Block A touches 6 blocks in total: 2 laterally placed, 3 below and 1 above it.

14. A
It is worth mentioning that the blocks are identical.
Block B touches 6 blocks in total: 3 laterally placed (2 vertical and 1 horizontal), and 3 are behind it.

15. D
Block C touches 2 blocks in total: 2 laterally placed, and another block below it.

Blocks

1. How many cubes are there in the figure?

   a. 30
   b. 32
   c. 35
   d. 24

2. How many cubes are there in the figure?

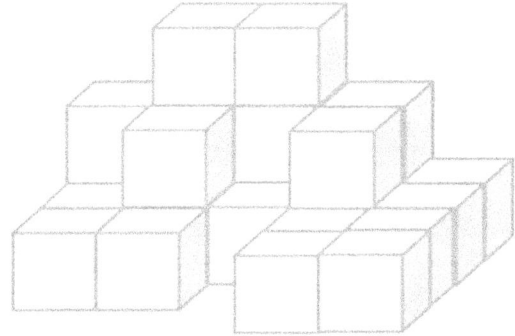

a. 30
b. 22
c. 15
d. 24

3. How many cubes must we add in the figure to form a perfect cube?

a. 55
b. 45
c. 70
d. 125

4. Which shape must we place to the right side of the figure to balance the weight of the system? All shapes are made by the same material.

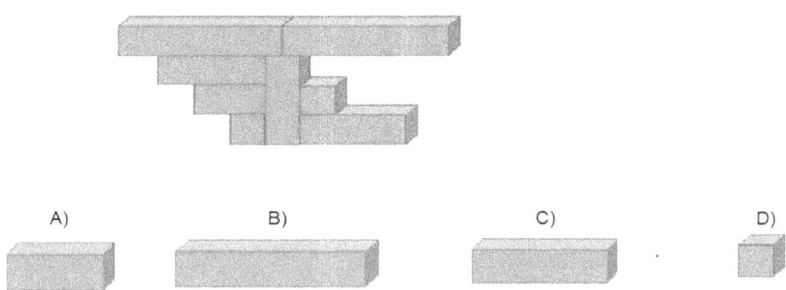

5. Which shape must we place on the existing figure to form a perfect square?

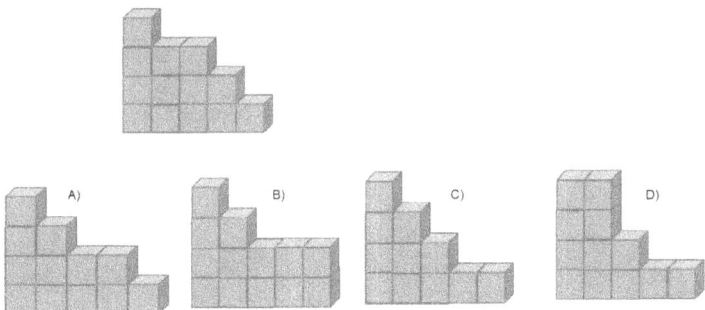

6. How many small cubes are missing in the figure to form a large perfect cube?

   a. 6
   b. 7
   c. 8
   d. 9

7. Which shape must we place on the existing figure to form a perfect cube?

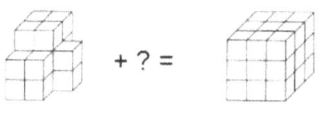

A) 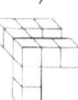 B)   C)   D)

8. Which shape must we place on the existing figure to form a perfect cube?

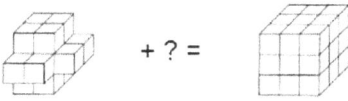

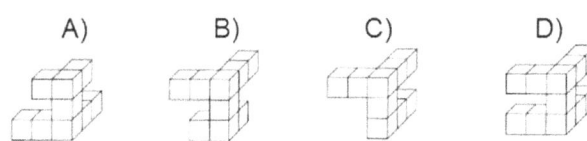

9. Which shape must we place on the existing figure to form a perfect cube?

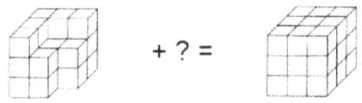

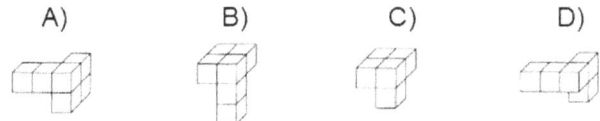

10. Which shape must we place on the existing figure to form a perfect cube?

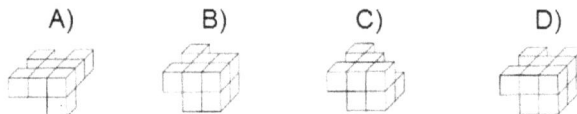

A)   B)   C)   D)

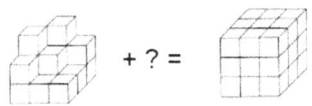

Answer Key

1. C
In the bottom row, there are 3 × 4 + 3 = 15 cubes.
In the next row, there are 2 + 3 + 3 + 3 + 1 = 12 cubes.
In the third row, there are 2 + 2 = 4 cubes.
In the upper row, there are only 2 cubes.
Thus, in total there are 15 + 12 + 4 + 2 = 33 cubes.

2. C
From the figure, you will see that there are 4 + 4 + 2 + 3 + 3 = 16 cubes in the bottom row, 2 + 1 + 2 + 1 = 6 cubes in the middle row and only 2 cubes in the upper row.

Thus, in total there are 16 + 6 + 2 = 24 cubes in the figure.

3. C
Since there are 5 cubes in the longest row, we need 5 × 5 × 5 = 125 cubes in total to form a perfect cube.

First, let's count the existing cubes. In the first row, there are 5 + 4 + 5 + 5 + 4 = 23 cubes.

In the second row, there are 5 + 3 + 4 + 4 + 1 = 17 cubes.
In the third row, there are 3 + 3 + 2 = 8 cubes.
In the fourth row, there are 3 + 3 + 1 = 7 cubes.

Thus, in total there are 23 + 17 + 8 + 7 = 55 cubes.

Hence, we must add 125 − 55 = 70 cubes to form a perfect big cube.

4. A
There are 4 kinds of shapes that are on the lateral sides of the vertical bar. These shapes vary in length. You may notice that on the right side, the second shortest shape is missing. Thus, it must be placed in that part to balance the system.

5. A

The existing figure has 4 rows and 5 columns, where not all of them are complete. To be a perfect square, it must have 5 × 5 dimensions.

From the figure, the first row is complete. In the second row one cube is missing, in the third row two cubes are missing, in the fourth row four cubes are missing and in the fifth row all 5 cubes are missing. There missing cubes must be filled with one of the shapes.

The only shape that fits the description is the first one.

6. B

A large perfect cube is formed when it has the dimensions 4 × 4 × 4.

The first row is complete, the second row has one cube missing, the third row has another cube is missing, and the fourth, upper row, 1 + 3 + 0 + 1 = 5 cubes are missing.

Hence, in total, 1 + 1 + 5 = 7 cubes are missing to form a perfect cube.

7. A

If you rotate the shapes in the choices by 90^0 clockwise, you will notice that the missing shape to form a perfect cube is the first one.

8. D

There are 5 missing cubes in the first row, 1 in the second and 5 in the third row. The only shape that fits the description is the fourth one. No rotation is needed.

9. B

There is one missing small square in the first row, one in the second and four squares in the third row. The only shape that fits the description is the second one. It does not need any rotation.

10. D
The first row does not have any missing cube. In the second row, only 3 cubes are missing at the closest corner. In the third row, 7 cubes are missing.

Cut Outs

1. Which figure is formed by assembling the following pieces?

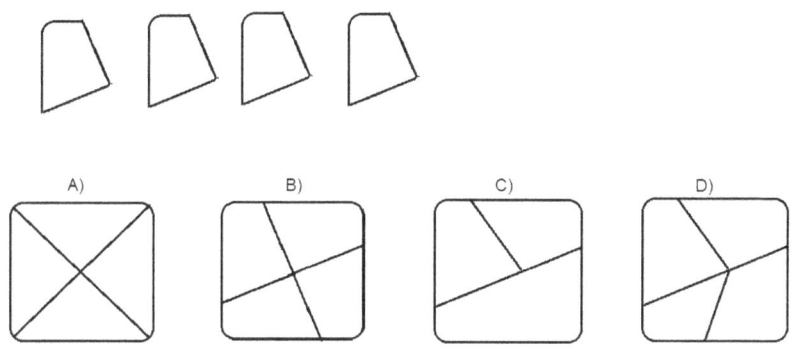

2. Which figure is formed by assembling the following pieces?

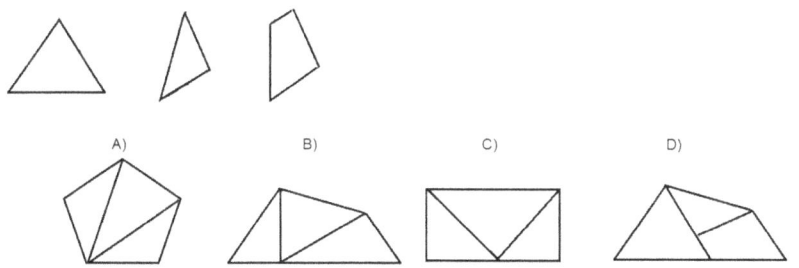

3. Which figure is formed by assembling the following pieces?

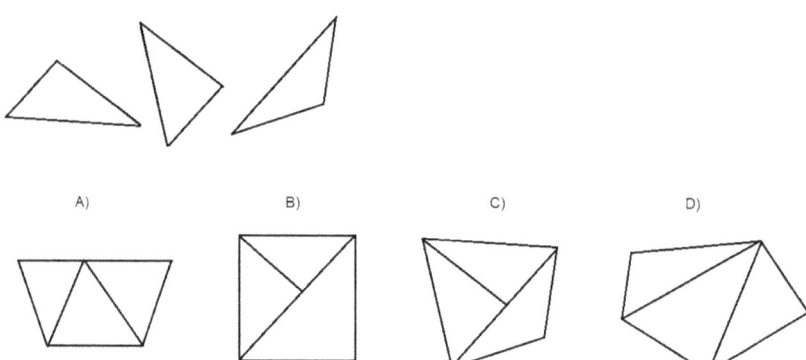

4. Which figure is formed by assembling the following pieces?

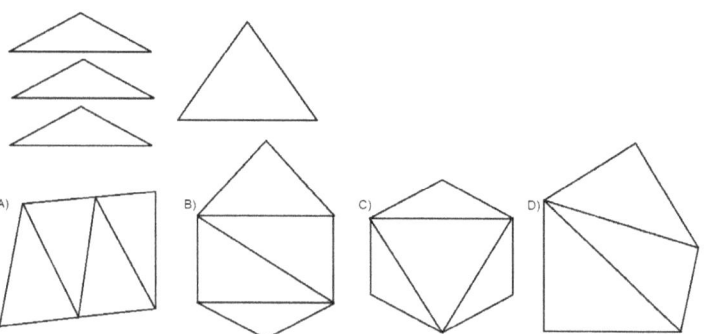

5. Which figure is formed by assembling the following pieces?

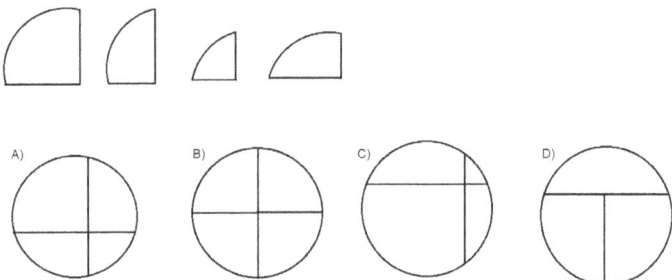

6. Which figure is formed by assembling the following pieces?

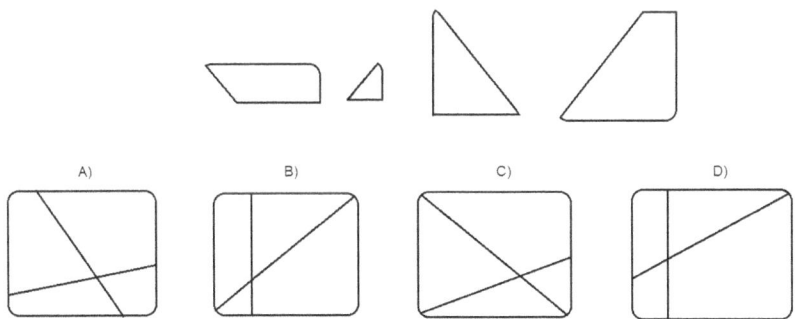

7. Which figure is formed by assembling the following pieces?

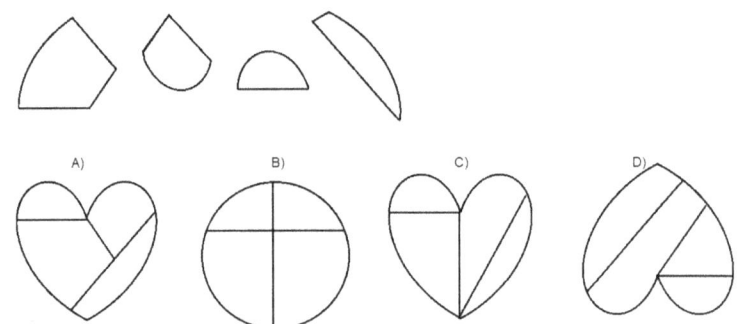

8. Which figure is formed by assembling the following pieces?

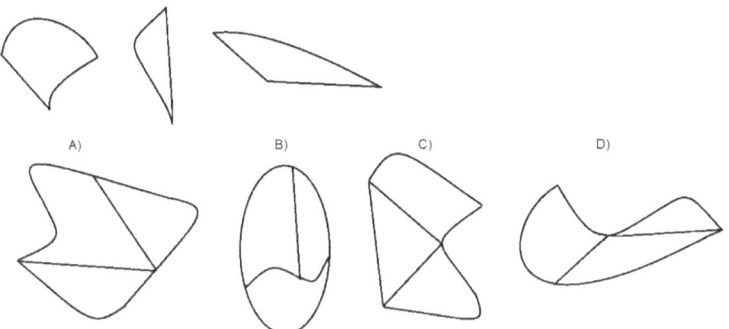

9. Which figure is formed by assembling the following pieces?

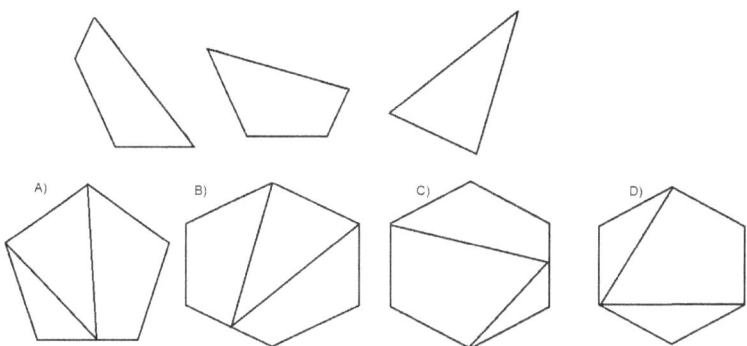

10. Which figure is formed by assembling the following pieces?

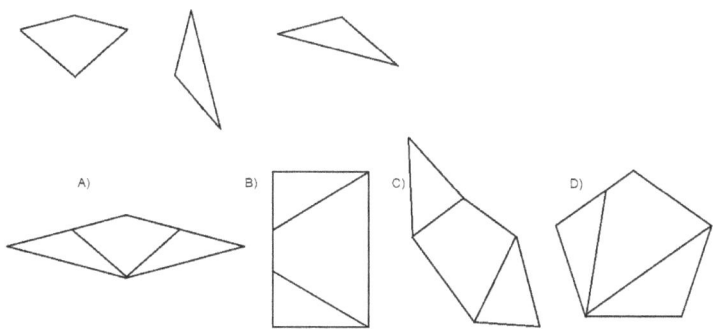

11. Which figure is formed by assembling the following pieces?

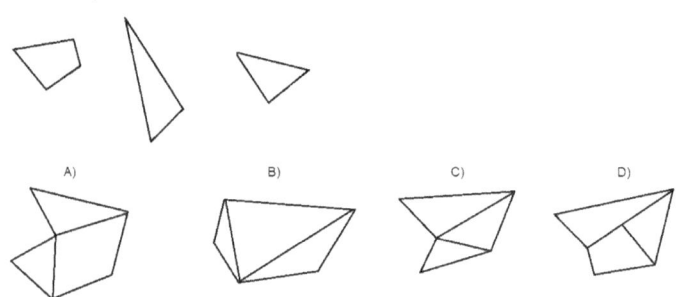

12. Which figure is formed by assembling the following pieces?

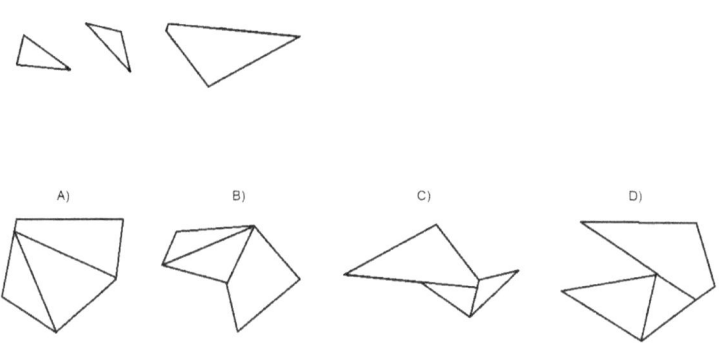

13. Which figure is formed by assembling the following pieces?

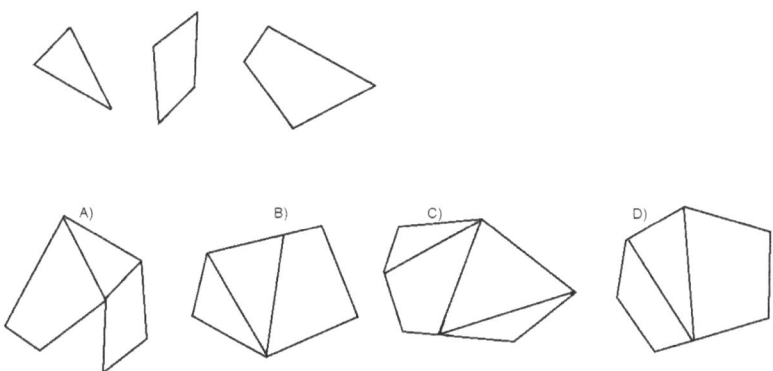

14. Which figure is formed by assembling the following pieces?

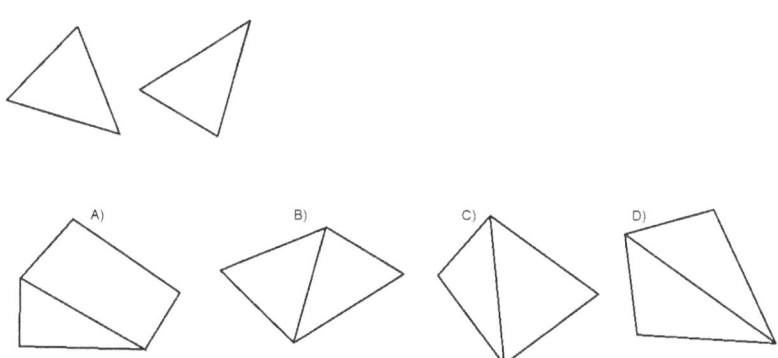

15. Which figure is formed by assembling the following pieces?

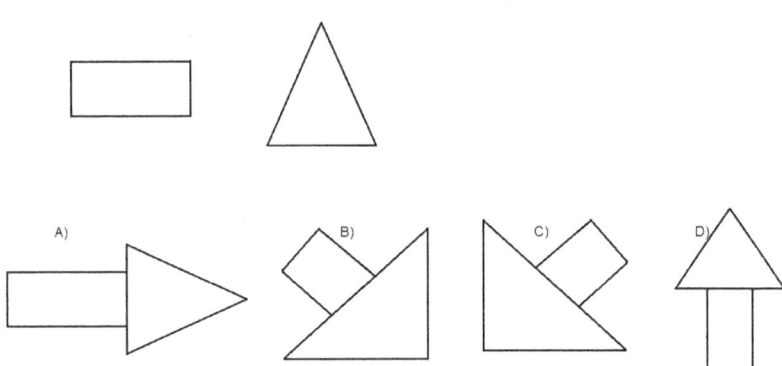

Answer Key

1. B
Since the pieces are identical, only the first two figures can qualify as candidates for the correct answer. The cutting traces in first figure represent the diagonals of the shape. This is not the case here. Hence, only the second figure fits the description.

2. D
There are two triangles and one quadrilateral in the separate pieces. Thus, since all figures except the fourth one contain only triangles, the choice D is the only that fits the description.

3. C
There are two right triangles and one wide triangle in the separate pieces. Only the third figure contains a wide triangle.

4. C
There are three identical right triangles and one equilateral triangle in the separate pieces. Only the third figure contains an equilateral triangle.

5. A
The figure B is excluded as it contains identical pieces. The fourth figure is excluded as well, as it contain only three pieces. So, we have to consider only A and C. The option C contains a very small and a very large piece that are not in the original figure.

6. B
There are two quadrilaterals and two triangles in the original figure. Choice A contains four quadrilaterals, thus it is not the right shape. Choice C contains one quadrilateral and three triangles, and choice D contains three quadrilaterals and one triangle. They are both incorrect.
The only remaining choice is the second one, which contains two quadrilaterals and two triangles.

7. A
Rotating the first piece by 900 anticlockwise, flipping the second piece vertically, and rotating the fourth piece by 90^0 clockwise, you will obtain the shape shown in the choice A.

8. D
Rotating the first piece by 900 anticlockwise, the second piece by 900 clockwise and the flipping the third piece vertically, you will obtain the figure shown in the choice D.

9. B
Rotating the first and the second shapes by 900 clockwise, and the third shape by 1800, you will obtain the figure shown in the choice B.

10. A
The first and the third shape do not need any rotation while the second shape rotates by 900 clockwise. As a result, you will obtain the figure shown in the first option.

11. D
The first shape rotates by 180^0, while the second and the third shape by 900 clockwise. As a result, you obtain the fourth figure.

12. C
The first and the third shape rotate by 1800, while the second shape rotates by 900 clockwise. As a result, you will obtain choice C.

13. C
The first and the second shape rotate by 180^0, while the third shape rotates by 900 anticlockwise.

14. B
Judging by the dimensions and the type of triangles, it is easy to conclude that choice B is the only one that fits the description.

15. A
Judging by the dimensions and the type of shapes, it is easy to conclude that the choice A is the only one that fits the description.

Jigsaw

	A	B	C	D
1	○	○	○	○
2	○	○	○	○
3	○	○	○	○
4	○	○	○	○
5	○	○	○	○
6	○	○	○	○
7	○	○	○	○
8	○	○	○	○
9	○	○	○	○
10	○	○	○	○
11	○	○	○	○
12	○	○	○	○
13	○	○	○	○
14	○	○	○	○
15	○	○	○	○

1. Which option completes the figure below?

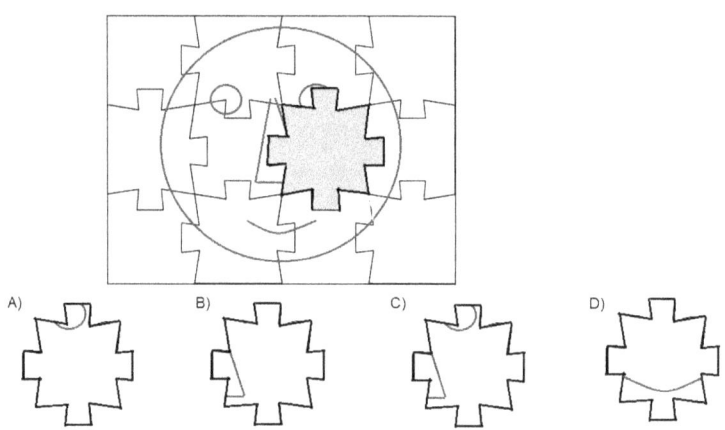

2. Which option completes the figure below?

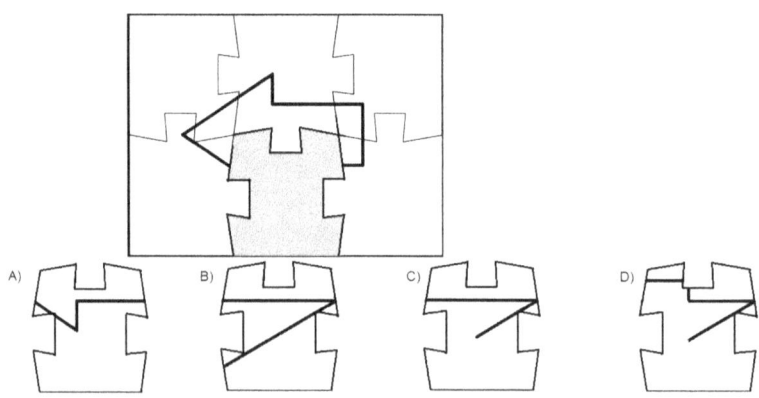

3. Which option completes the figure below?

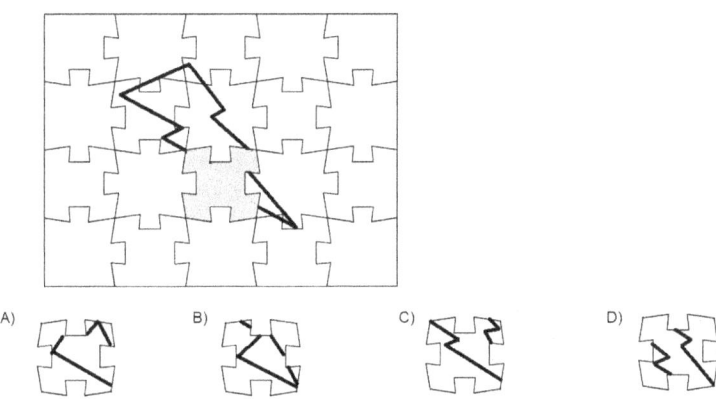

4. Which option completes the figure below?

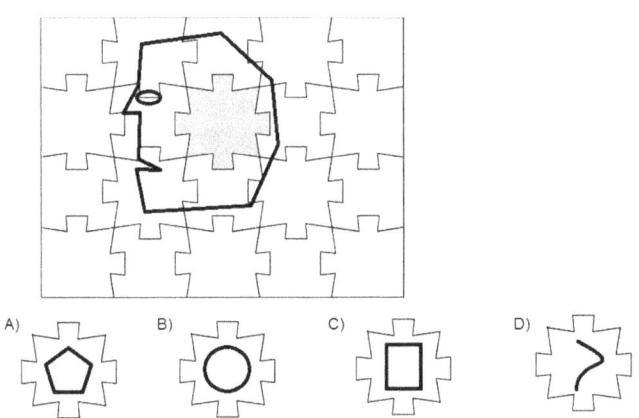

5. Which option completes the figure below?

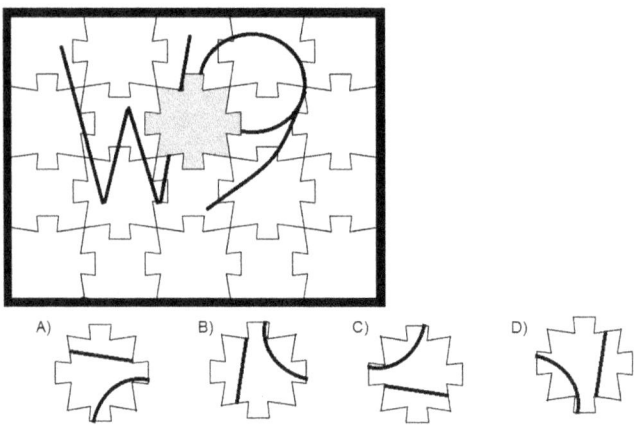

6. Which option completes the figure below?

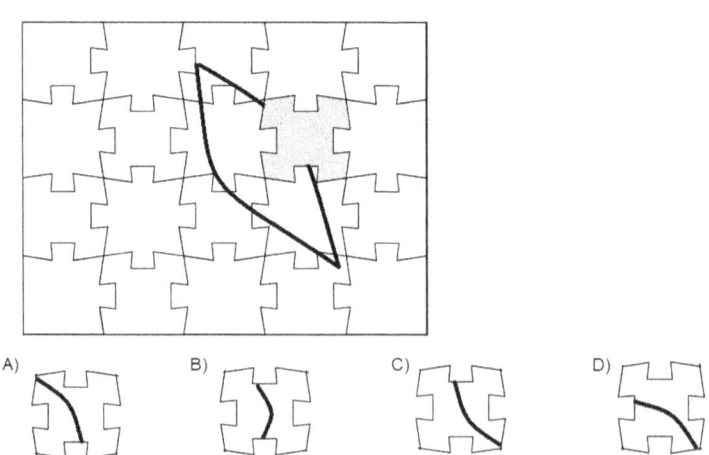

7. Which option completes the figure below?

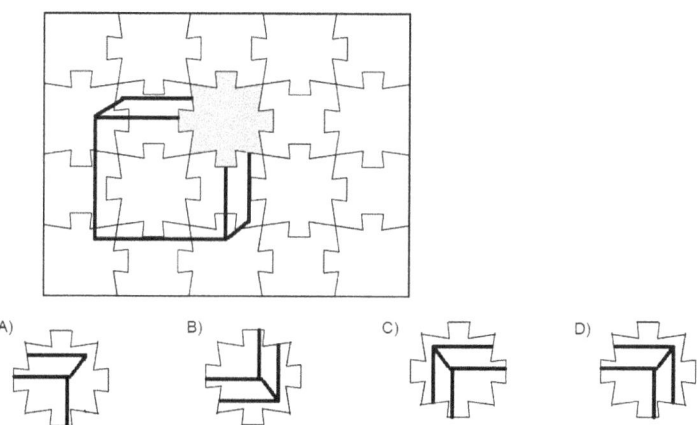

8. Which option completes the figure below?

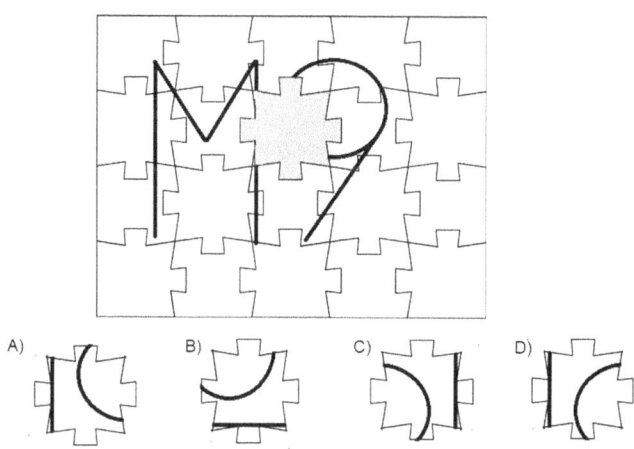

9. Which option completes the figure below?

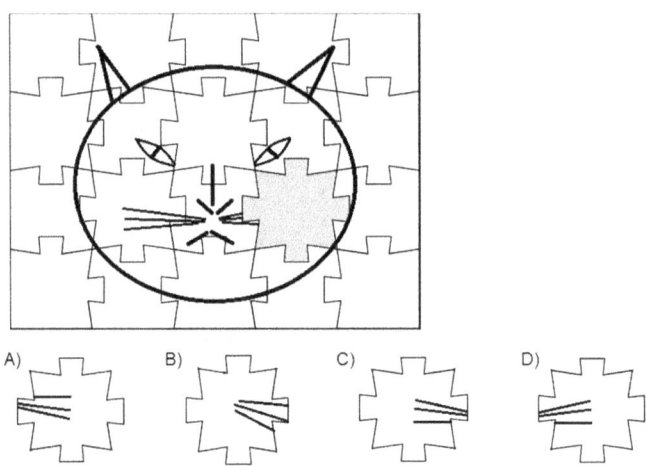

10. Which option completes the figure below?

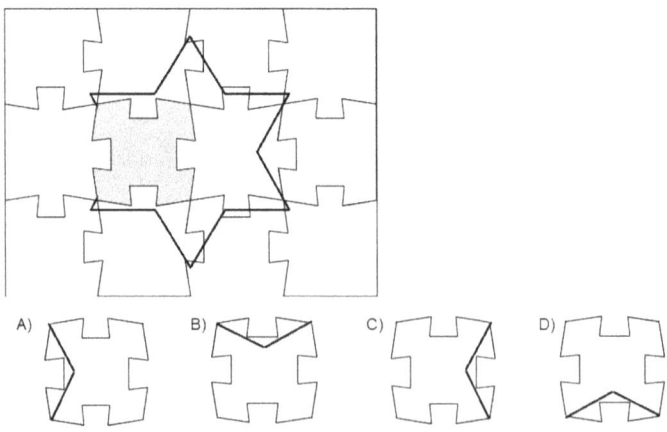

11. Which option completes the figure below?

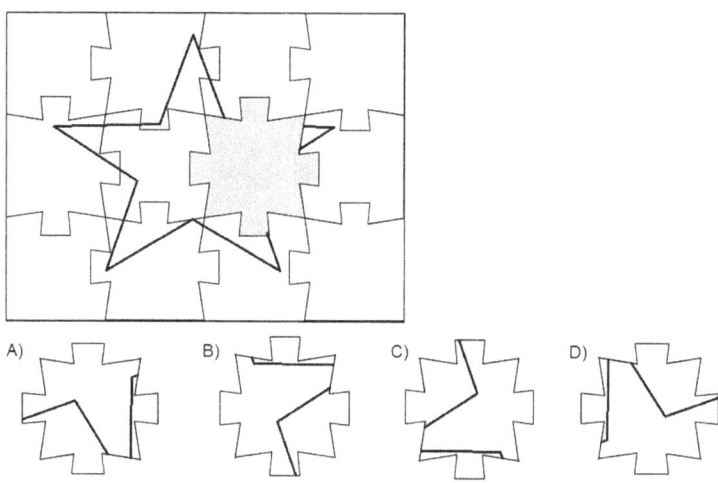

12. Which option completes the figure below?

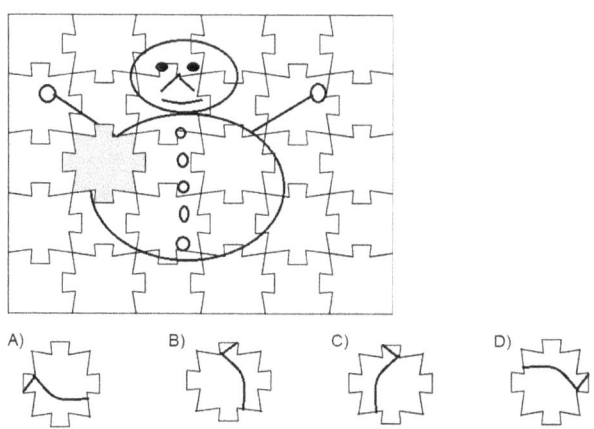

13. Which option completes the figure below?

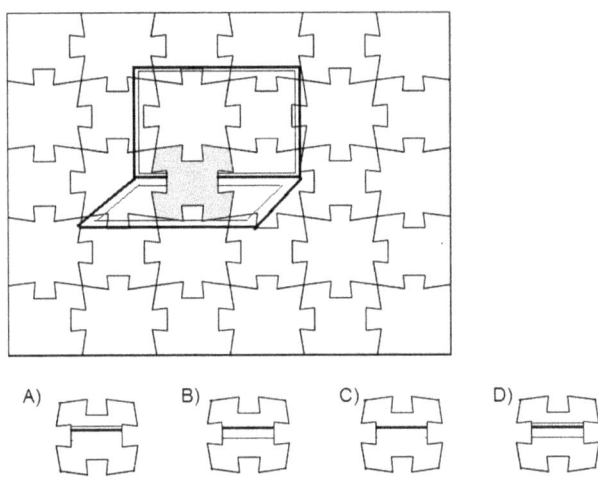

14. Which option completes the figure below?

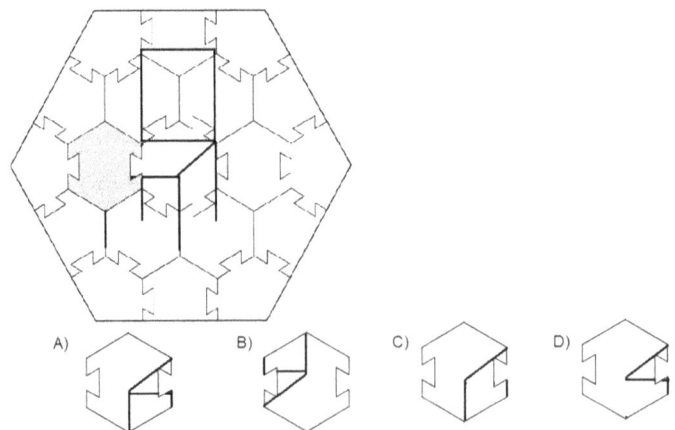

15. Which option completes the figure below?

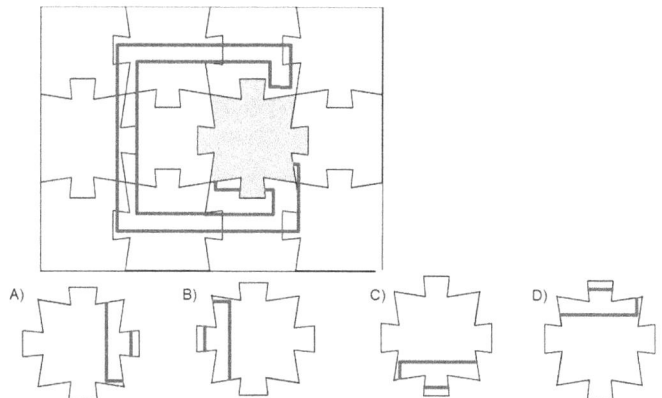

Answer Key

1. C
The figure represents a symmetrical face, with part of the eye and a part of the nose are missing.

The only piece that fits the description is choice C.

2. A
The figure represents a kind of symmetrical horizontal arrow. Therefore, a part of the lateral edge of the tip is missing.

The only piece that fits the description is choice A.

3. C
The figure represents the lightning symbol. You can see that the line interrupts at the upper left corner of the missing piece, so it must continue in that place.

The only piece that fits the description is choice C.

4. D
The figure represents a kind of human head in the lateral position. You can imagine that in the place of missing piece there must be an ear.

The only piece that fits the description is choice D, as the figure is the closest to the ear shape.

5. B
The figure represents a kind of script, more precisely two letters: W and 9. Thus, in the missing piece, there must be a part of W (a line) and a part of the number 9 (a curve).

The only shape that fits the description is choice B.

6. A
The figure represents a kind of tree-leaf-shaped object.

Judging by the leaf symmetry, the missing piece is in choice A.

7. A
The figure represents a 3-D cube. Judging by the cube's symmetry, the missing piece represents the upper-right corner.

8. A
The figure represents a kind of script, more precisely two letters: M and 9. The missing piece, must contain a part of M (a vertical line) and a part of the number 9 (the lower-left part of a circle).

The only shape that fits the description is choice A.

9. D
The figure represents a cat face. The missing piece must contain the right part of the whiskers. Judging by the symmetry of the figure, the whiskers are pointed up (left to right).

The only figure that fits the description is choice D.

10. A
The figure represents a six-pointed star. Judging by the figure's symmetry, choice A represents the missing piece.

11. B
The figure represents a pentacle. The missing piece is the one shown in choice B.

12. C
The figure represents a kind of snowman shape. Thus, by symmetry, the missing piece is choice C.

13. D
The figure resembles to a laptop. The thick line shows the laptop borders and the thin lines represent the screen border and the keyboard border respectively.

The missing shape must contain all of them, choice D.

14. A
The figure resembles to a chair in which the middle left part is missing.

15. C

The figure resembles the letter G with the left part is missing.

Choice C represents the missing piece.

Matching Shapes

	A	B	C	D
1	○	○	○	○
2	○	○	○	○
3	○	○	○	○
4	○	○	○	○
5	○	○	○	○
6	○	○	○	○
7	○	○	○	○
8	○	○	○	○
9	○	○	○	○
10	○	○	○	○
11	○	○	○	○
12	○	○	○	○
13	○	○	○	○
14	○	○	○	○
15	○	○	○	○

Questions 1 - 4 refer to the following figure

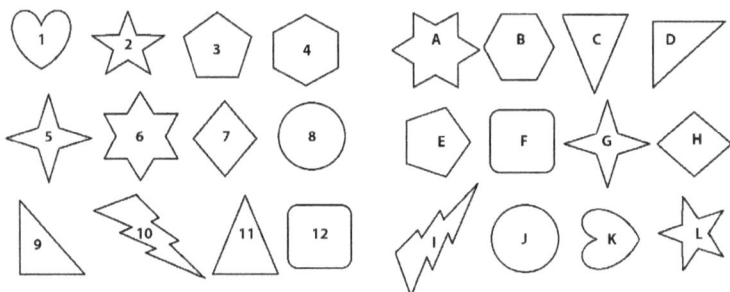

1. Which of the following set of matching pairs are correct?

    a. 1-f   3-g   10-k
    b. 7-c   9-d   1-j
    c. 4-h   11-c   9-d
    d. 8-j   5-g   9-d

2. Which of the following set of matching pairs are correct?

    a. 11-d   6-l   4-e
    b. 12-j   5-l   7-e
    c. 9-d   11-c   6-a
    d. 9-d   2-l   6-l

3. Which of the following set of matching pairs are correct?

a. 3-1   5-g   2-k
b. 7-h   3-e   2-l
c. 6-a   11-d 12-f
d. 10-i 5-g   4-b

4. Which of the following set of matching pairs are correct?

a. 7-f   9-c   12-f
b. 5-l   7-g   9-d
c. 1-j   12-f 6-l
d. 5-g   11-c 1-k

Questions 5 - 7 refer to the following figure

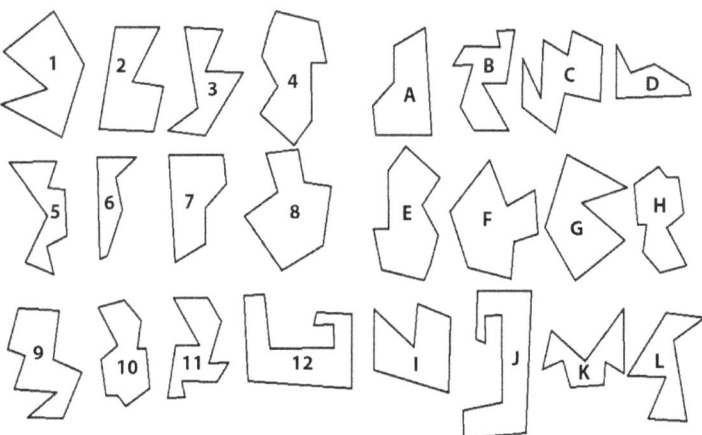

5. Which of the following set of matching pairs are correct?

   a. 2-k   7-l   5-b
   b. 11-b   4-e   2-i
   c. 8-g   6-d   10-k
   d. 12j   9-c   3-h

6. Which of the following set of matching pairs are correct?

   a. 1-g   12-j   8-f
   b. 9-c   1-g   5-l
   c. 4-e   5-k   7-c
   d. 6-d   10-a   11-c

7. Which of the following set of matching pairs are correct?

   a. 10-h   8-f   4-l
   b. 10-e   5-d   11-b
   c. 6-d   10-h   12-j
   d. 5-g   7-I   10-a

Questions 8 - 11 refer to the following figure

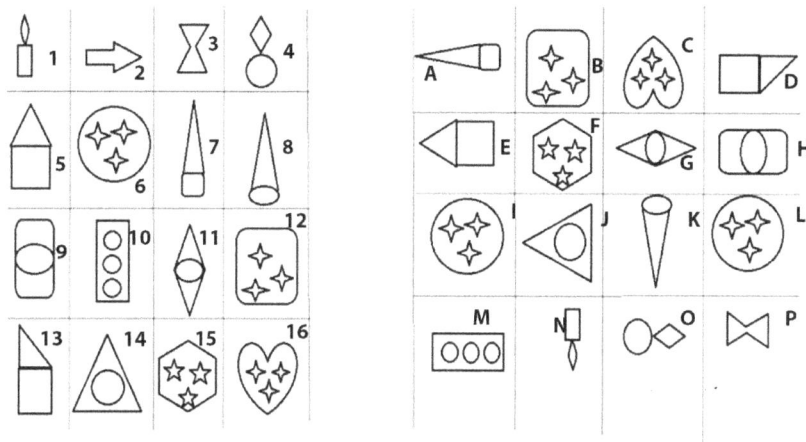

8. Which of the following statements about the shapes in the figure below is correct?

    a. 1, 7, 10 and 16 match with m, j, k, and p respectively

    b. 6, 7, 8 and 13 match with i, a, k and d respectively

    c. 2, 5, 9 and 10 match with b, f, d and e respectively

    d. 4, 10, 11 and 15 match with m, n, o and p respectively

9. Which of the following statements about the shapes in the figure below is correct?

    a. 1, 7, 10 and 16 match with n, a, m, and c respectively

    b. 2, 6, 8 and 12 match with i, a, k and d respectively

    c. 2, 5, 9 and 10 match with c, f, k and e respectively

    d. 4, 10, 11 and 15 match with m, n, o and b respectively

10. Which of the following statements about the shapes in the figure below is correct?

   a. 5, 7, 9 and 11 match with e, g, i and l respectively
   b. 1, 2, 3 and 4 match with a, b, c and d respectively
   c. 13, 14, 15 and 16 match with m, n, o and p respectively
   d. 9, 10, 11 and 12 match with h, m, g and b respectively

11. Which of the following statements about the shapes in the figure below are incorrect?

   a. All shapes on the left part of the figure have a single match on the right part of the figure
   b. The upper row of numbered part of the table corresponds to the second column of the lettered part of the table
   c. 14 corresponds to j but 5 does not correspond to d
   d. The shape n is obtained by turning the shape 1 upside down

Questions 12 - 15 refer to the following figure

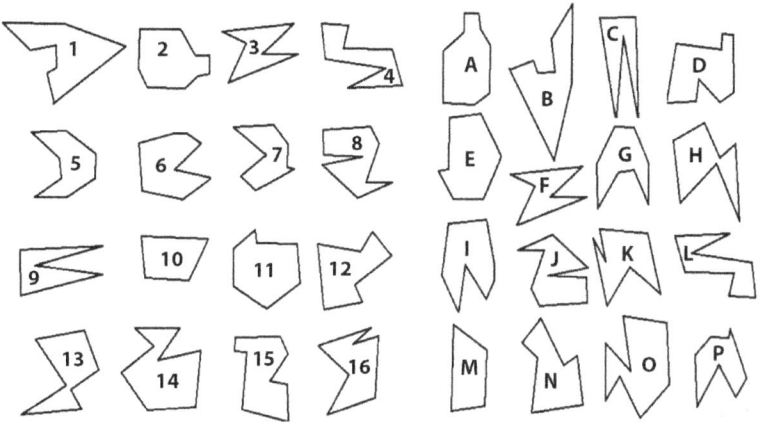

12. Which of the following statements about the shapes in the figure below is correct?

    a. 1, 3, 5 and 16 match with n, a, m, and c respectively
    b. 2, 6, 7 and 12 match with i, a, k and d respectively
    c. 4, 5, 9 and 10 match with c, f, k and e respectively
    d. 5, 6, 9 and 11, match with g, i, c and e respectively

13. Which of the following statements about the shapes in the figure below is correct?

    a. 1, 2, 5 and 6 match with a, b, c and d respectively
    b. 13, 14, 15 and 16 match with d, e, f and m respectively
    c. 2, 10, 11 and 12 match with a, m, e and n respectively
    d. 6, 8, 9 and 10 match with c, f, g and o respectively

14. **Which of the following statements about the shapes in the figure below is correct?**

    a. 9 matches with f but 15 doesn't match with d
    b. 1 matches with b but 15 doesn't match with e
    c. 14 matches with k but 7 doesn't match with g
    d. 10 matches with m but 11 doesn't match with e

15. **Which of the following pairs of shapes do NOT match?**

    a. 9-c   13-h   16-k
    b. 2-a   6-i   14-o
    c. 13-h   15-d   16-k
    d. 9-c   6-I   4-k

Answer Key

1. D
Choice D is correct as 8 and j both represent circles, 5 and g both represents four pointed stars, while 9 and d both represent right triangles.

2. C
Choice C is correct as 9 and d both represent right triangles, 11 and c both represent isosceles triangles while 6 and both represent six-pointed stars.

3. B
Choice B is correct as 7 and h both represent rhombuses, 3 and e both represent pentagons while 2 and l both represent pentacles.

4. D
Choice D is correct as 5 and g both represent four pointed stars, 11 and c both represent isosceles triangles while 1 and k both represent heart-like shapes.

5. B
Choice B is correct as b is obtained by the 180^0 rotation of 11, e is obtained by rotating 4 by 180^0 and i is obtained by a 90^0 anticlockwise rotation of 2.

6. A
Choice A is correct as g is obtained by rotating 1 by 180^0, j is obtained by rotating 12 by 90^0 anticlockwise and f is obtained by rotating 8 by 90^0 clockwise.

7. C
Choice C is correct as d is obtained by rotating 6 by 90^0 anticlockwise, h is obtained by rotating 10 by 180^0 and j is obtained by rotating the shape 12 by 90^0 anticlockwise.

8. B
6 matches with i, 7 with a, 8 with k and 13 with d, regardless of the figures' rotation.

9. A
1 matches with n, 7 with a, 10 with m and 16 with c, regardless of the figures' rotation.

10. D
9 matches with h, 10 with m, 11 with g and 12 with b, regardless of the figures' rotation.

11. B
Choice A is correct. Each shape in the left part of the figure has a unique match in the right part.
Choice B is incorrect. The upper row of numbered part of the table does not correspond to the second column of the lettered part of the table.
Choice C is correct. 14 corresponds to j but 5 corresponds to e, not d.
Choice D is correct. The shape n is obtained by turning the shape 1 upside down.

12. D
5 matches with g, 6 with i, 9 with c and 11 with e, regardless of the figures' rotation.

13. C
2 matches with a, 10 with m, 11 with e and 12 with n, regardless of the figures' rotation.

14. B
Choice A is incorrect. 9 does not match with f.
Choice B is correct. 1 matches with b but 15 doesn't match with e.
Choice C is incorrect. 14 does not match with k.
Choice D is incorrect because 11 matches with e.

15. D
All choices match except choice D, 6-l and 4-k.

Visual Comparison

	A	B	C	D
1	○	○	○	○
2	○	○	○	○
3	○	○	○	○
4	○	○	○	○
5	○	○	○	○
6	○	○	○	○
7	○	○	○	○
8	○	○	○	○
9	○	○	○	○
10	○	○	○	○
11	○	○	○	○
12	○	○	○	○
13	○	○	○	○
14	○	○	○	○
15	○	○	○	○

1. Which figure has the greatest shaded area?

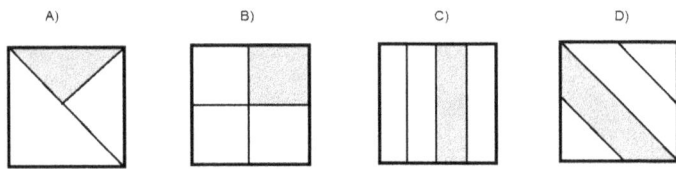

2. Which figure need less color to paint?

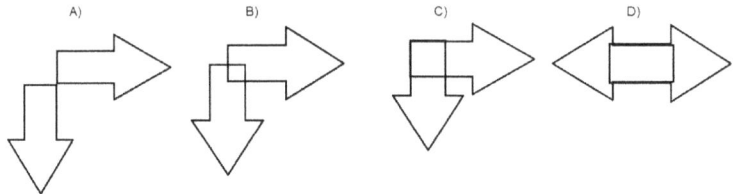

3. Which figure has the smallest shaded area?

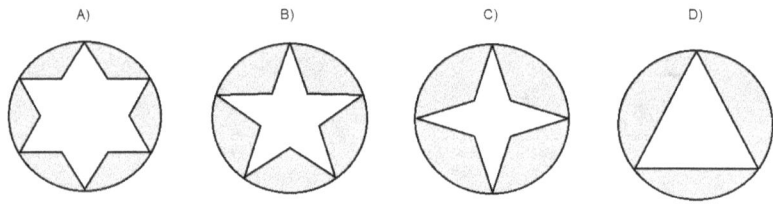

4. Which figure has the smallest shaded area?

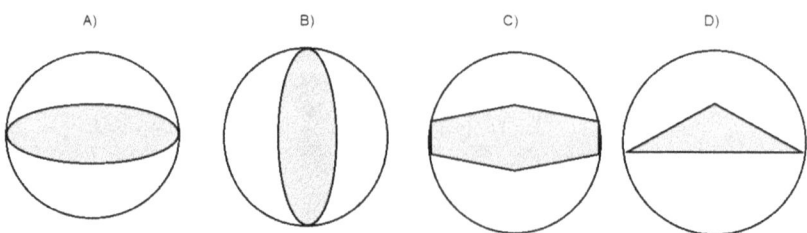

5. Which gear requires more material to produce?

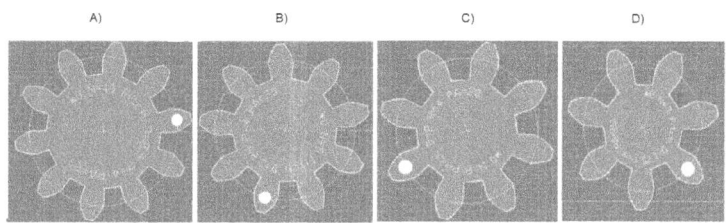

6. Which triangle will hold the largest inscribed circle?

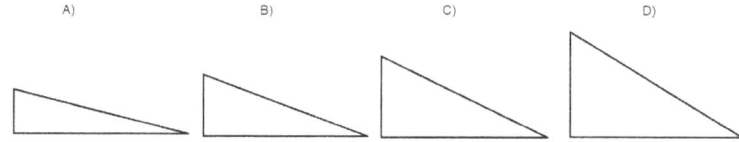

7. Which figure has the greatest space between the two shapes?

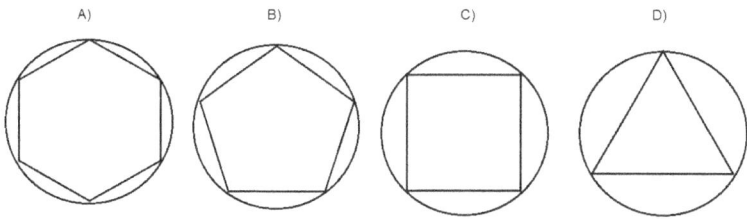

8. Which figure will hold the largest inscribed circle?

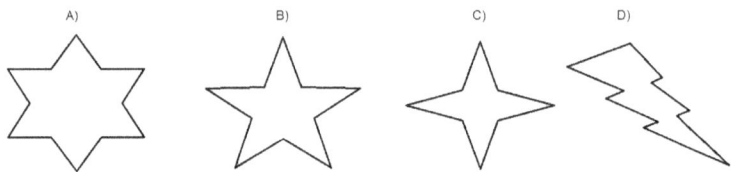

9. Which figure below will take the greatest amount of wire?

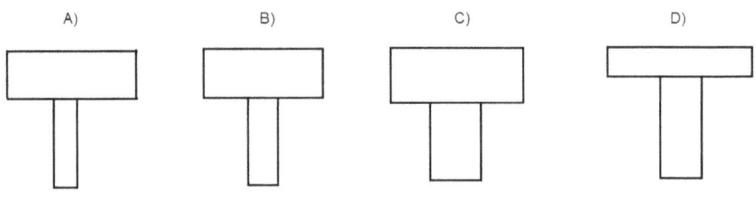

10. Which of the shapes below is the smallest?

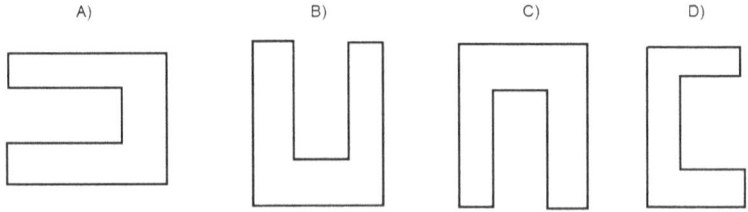

11. Which flower below is the biggest?

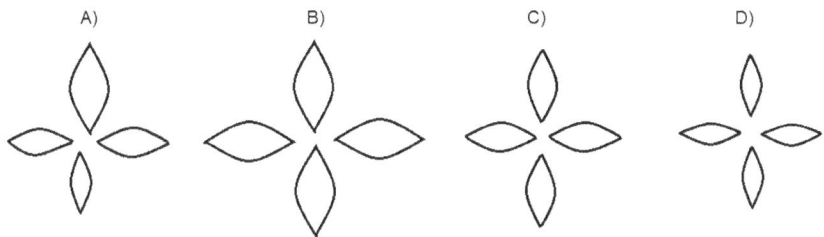

12. If the shapes shown below are made of metal, which shape uses a greater amount of metal?

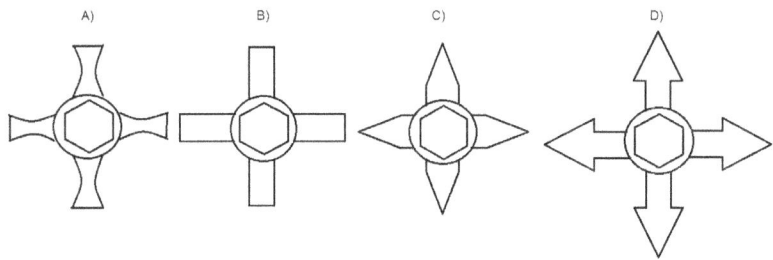

13. Which shape below uses the most paint?

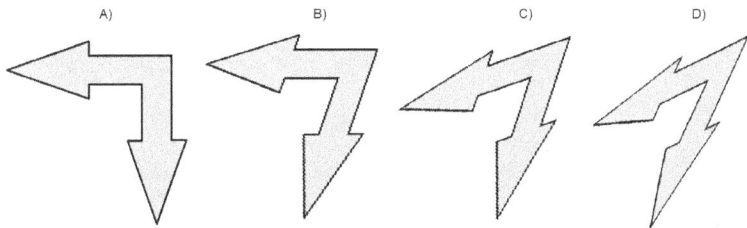

14. Which figure will hold the largest inscribed circle?

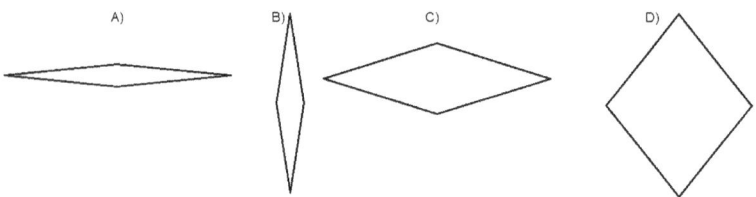

15. Which of the inscribed figures below is the smallest?

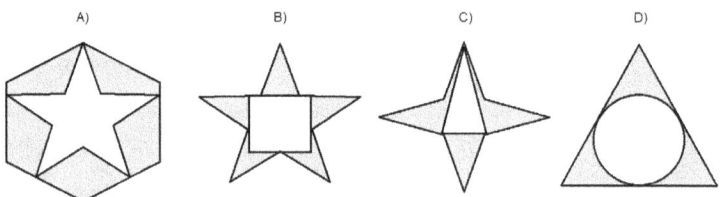

VISUAL COMPARISON

Answer Key

1. D
In the first three figures, one quarter is shaded. Only choice D has a shaded area greater than one quarter.

2. D
The figure that needs the least paint is the one with the smallest surface area. In choice D, the arrows overlap the most, so the surface area is the smallest and requires the least paint.

3. A
Figure A has the most white space. Since all the circles are identical, the shaded area of the first figure is the smallest.

4. D
Choice D has more white space than the other shapes. Since all the circles are identical, the shaded area of choice D (the triangle) is the smallest.

5. A
Choice A has a greater internal diameter (without including the teeth), and has more teeth, and therefore requires more material to produce.

6. D
Choice D is the largest triangle and will hold the largest inscribed circle.

7. D
Choice D has more space between the inner shape (triangle) and the outer shape (circle).

8. A
Choice A has the largest interior volume and can contain the largest inscribed circle.

9. C
The larger the rectangles that make up the figure, the more wire will be required. Choice C has the largest area, and requires the most wire to produce.

10. D
The first three shapes have one short and two long sides. Choice D is the smallest, with one long and two short sides.

11. B
Choice A has 1 small, 2 medium, and one large petal.
Choice B is the largest, with 4 large petals.
Choice C has 4 medium size petals.
Choice D has 4 small petals.

12. D
Looking carefully at the dimensions and thickness of the figures, choice D is the largest and will take the most material to produce.

13. A
When the angle between the arrows decreases, their thickness also decreases. So the largest dimension (choice A) will be when the angle is 90^0.

14. D
The wider the figure, the larger the inscribed circle. Therefore, choice A will contain the largest inscribed circle.

15. C
Choice C is correct. Choice B and C are very similar, however, since for similar dimensions a triangle, has a smaller area than a rectangle, the hesitation perishes.

Conclusion

CONGRATULATIONS! You have made it this far because you have applied yourself diligently to practicing for the exam and no doubt improved your potential score considerably! Getting into a good school is a huge step in a journey that might be challenging at times but will be many times more rewarding and fulfilling. That is why being prepared is so important.

Good Luck!

Register for Free Updates and More Practice Test Questions

Register your purchase at

https://www.test-preparation.ca/register/ for updates and free test tips and more practice test questions.

Online Resources

How to Prepare for a Test - The Ultimate Guide

https://www.test-preparation.ca/prepare-test/

Learning Styles - The Complete Guide

https://www.test-preparation.ca/learning-style/

Test Anxiety Secrets!

https://www.test-preparation.ca/test-anxiety/

Time Management on a Test

https://www.test-preparation.ca/time-management/

Flash Cards - The Complete Guide

https://www.test-preparation.ca/flash-cards/

Test Preparation Video Series

https://www.test-preparation.ca/test-video/

How to Memorize - The Complete Guide

https://www.test-preparation.ca/memorize/

www.ingramcontent.com/pod-product-compliance
Lightning Source LLC
Chambersburg PA
CBHW072000070526
44583CB00015B/1272